Autobiographical Memory and the Covid-19 Pandemic

Malen Migueles Seco • Alaitz Aizpurua Sanz
Editors

Autobiographical Memory and the Covid-19 Pandemic

Springer

Editors
Malen Migueles Seco
Faculty of Psychology
University of the Basque Country (UPV/EHU)
Donostia-San Sebastian, Spain

Alaitz Aizpurua Sanz
Faculty of Psychology
University of the Basque Country (UPV/EHU)
Donostia-San Sebastián, Spain

ISBN 978-3-032-03894-4 ISBN 978-3-032-03895-1 (eBook)
https://doi.org/10.1007/978-3-032-03895-1

This Springer imprint is published by the registered company Springer Nature Switzerland AG
The registered company address is: Gewerbestrasse 11, 6330 Cham, Switzerland

Prologue

In 2020, the COVID-19 pandemic transformed our routines, relationships and priorities, but it also altered the way we remember. Our memories—both the most emotional and the seemingly trivial—were recorded with intensity and, at the same time, with a disconcerting fragility. Reflecting on the pandemic evokes images and anecdotes. During the confinement and the first months of the pandemic, we watch on television the deserted streets of bustling cities, the chaos in hospitals and the rising numbers of infected and dead, and one phrase came to mind: 'This too shall pass'. It was a phrase of consolation, but it also reflected the conviction that we were living through a unique period, with shadows and lights, that we would remember many times. And, in order to remember, we need memory with its multiple dimensions and limitations.

This collective volume examines, through the prism of memory, the complex period of the COVID-19 pandemic. It brings together seven chapters by researchers from diverse disciplines and countries, exploring how a global, unexpected and prolonged event has reshaped autobiographical memory from various perspectives: emotional, cognitive, social, cultural and narrative. In all the chapters, individual differences are analysed, showing that the pandemic behaved chameleon-like and affected people differently according to age, gender or personal conditions.

The tour begins with a review of the emotional and cognitive impact of COVID-19. Aranberri and Liberal offer a comprehensive review of the emotional state, psychological well-being and cognitive processes affected by the pandemic, reminding us that our emotions not only shape our memories but also determine the present and direct the future. This chapter documents how fear, anxiety and sadness marked the first months, particularly affecting young people and those with maladaptive coping strategies. It also reflects on protective factors such as gratitude, hope and social connection, which helped individuals resist and adapt to the most difficult moments of the crisis.

The next chapter, by Podlesek and Kavcic, explores the cognitive effects of the pandemic, both on those who contracted the disease and on those who endured confinement under uncertainty and stress. Their results show alterations mainly in attention, memory and decision-making. The authors are interested in both providing data on the neurological mechanisms involved in the observed impairments and highlighting factors such as the impact of chronic stress, isolation or emotional distress on cognitive functioning.

From a more social and cultural perspective, in the third chapter, Dönerkayalı, Watson and Öner provide a clear conceptual framework for understanding how the pandemic affected collective memory. They highlight the role of globally shared themes, such as lockdowns, infections and health-related anxieties, in showing how societies narratively reconstruct their common experiences of the pandemic. They explain how memory ceases to be a strictly individual phenomenon and takes on social, political and cultural dimensions. This perspective broadens the focus, showing not only how we remember together, but also how we forget together.

In the fourth chapter, Allé and Berntsen present spontaneous cognition, that is, the involuntary retrieval of autobiographical memories of the past, or the spontaneous imagination of future events. Through three studies, they examine this involuntary mental time travel during the COVID-19 pandemic, analysing its relationship to the pandemic context, conditions of confinement and individual mental health. Their data explore the increase in involuntary memories and future thoughts that proliferated during confinement and the months of the pandemic. These spontaneous, often intrusive, autobiographical episodes reflect the tension between the need to process lived experience and the difficulty of integrating such disturbing experiences into our personal narrative.

Lanciano and Curci analyse, in the fifth chapter, the so-called flashbulb memories, that is, vivid, detailed and confidently held memories of learning about emotionally significant events. During the pandemic, highly emotional initial moments, such as the announcement of the confinement, the appearance of the first infected person in the country or the start of the administration of effective vaccines triggered very vivid mental images of these events. These memories, intense and stable over time, reveal the emotional impact that certain collective events have on individual autobiography. The chapter reflects on whether a prolonged and changing event such as COVID-19 can also generate flashbulb memories, traditionally associated with surprising, specific and highly shocking events.

The chapter by Migueles and Aizpurua alerts us to the dangers of the disinformation and 'infodemia' that accompanied the health crisis, as factors that promoted an ideal breeding ground for the generation and proliferation of false memories. Thus, in an information-saturated environment, memory became particularly vulnerable to distortion, giving rise to false memories that affected behaviour and decision-making in those difficult times. Paradoxically, memory also served an adaptive function, biasing towards positive content, which, especially in older individuals, helped maintain personal balance and improved their sense of well-being.

The volume closes with a contribution by Ekinci and Brown, who reflect on how the pandemic has restructured our autobiographical memory. They start from transition theory, which argues that autobiographical memories are organised around significant life transitions or transformations (e.g., becoming independent, finishing university, moving to a new city) that act as turning points and temporal organisation. In this sense, the pandemic can be perceived as a shared milestone with remarkable effects on autobiographical memory. The chapter invites us to rethink the role of collective events as structuring autobiographical memory, opening new avenues for the study of memory in shared contexts.

As editors, it has been a privilege to accompany this process, to read the diverse voices that make up this work and to see how science, memory and experience can intertwine to make sense of what was lived. Far from offering closed answers, this book opens up paths and becomes an invitation to test our memory: How do we remember the pandemic? What have we learned from it? How has it changed us? How will we remember it in the future? In a fast-moving world, taking a moment to reflect on how we remember can be a way to heal, understand ourselves better and prepare for the challenges ahead. In times of uncertainty, remembering can be an act of resilience, as well as an exercise in reflection, care and responsibility.

Donostia-San Sebastián, Spain
July 2025

Malen Migueles Seco
Alaitz Aizpurua Sanz

Contents

Emotional State and Psychological Well-Being During COVID-19 Pandemic and Their Impact on Cognition

Ainara Aranberri and Unai Liberal

Abstract The COVID-19 pandemic constituted a disruptive global event that significantly impacted *emotional states*, *psychological well-being*, and *cognitive functioning* across diverse populations. This chapter analyzes these three interconnected domains through a systematic literature review conducted according to PRISMA guidelines. An initial search yielded 1404 articles across six major academic databases. After applying defined inclusion and exclusion criteria, a final sample of 40 empirical studies was selected for qualitative synthesis. The results are presented in three thematic categories. First, emotional state: the pandemic elicited intense negative emotions such as fear, sadness, anger, and anxiety, particularly during its early stages. Younger individuals, students, and those using maladaptive coping strategies were most affected. Over time, however, emotional trajectories showed signs of adaptation, with gratitude, hope, and emotional coherence emerging as protective emotional traits. Second, psychological well-being: multiple studies reported declines in well-being, particularly among adolescents, students, women, and vulnerable populations. Increases in depression, anxiety, loneliness, and suicidal ideation were common. However, certain groups—especially older adults—exhibited resilience and emotional stability. Predictors of well-being included age, gender, income, coping mechanisms, daily routines, and access to social support. Third, cognitive processes: the pandemic disrupted cognitive functions such as appraisal, attention, emotion regulation, memory, and decision-making. Maladaptive cognitive patterns, including catastrophic thinking and perceived helplessness, were linked to greater psychological distress. In contrast, adaptive strategies such as cognitive reappraisal, functional problem-solving, and positive social connections were associated with improved outcomes. Cognitive consequences also included memory distortions, spontaneous thought disruptions (e.g., mind-wandering, dreaming), and decision-making impairments. In conclusion, the pandemic generated a complex psychological impact. Emotional, well-being, and cognitive responses varied across

A. Aranberri (✉) · U. Liberal
Faculty of Psychology, University of the Basque Country (UPV/EHU), Donostia-San Sebastian, Spain
e-mail: ainara.aranberri@ehu.eus; unai.liberal@ehu.eus

M. Migueles Seco, A. Aizpurua Sanz (eds.), *Autobiographical Memory and the Covid-19 Pandemic*, https://doi.org/10.1007/978-3-032-03895-1_1

individuals and contexts, highlighting the need for integrated and personalized mental health strategies during and after collective crises.

Keywords COVID-19 · Emotional state · Psychological well-being · Cognition

1 Introduction

The COVID-19 pandemic has had a profound impact not only on physical health but also on individuals' emotional states, psychological well-being, and cognitive functioning. This chapter offers an in-depth analysis of these three interconnected domains during the pandemic, drawing on contemporary psychological theories and empirical evidence. Through a dual methodological approach—combining theoretical discussion and a systematic literature review following PRISMA guidelines—the chapter explores how emotional responses evolved, how well-being was challenged or maintained, and how cognitive processes were influenced in the face of prolonged uncertainty. The structure of the chapter is divided into conceptual sections addressing the definitions and theoretical foundations of emotional state, well-being, and cognition, followed by the methods and results of the systematic review. The results are presented in three main thematic blocks: emotional states, psychological well-being, and cognition. And the final discussion integrates these findings to highlight the complex psychological consequences of the pandemic.

Autobiographical memory is a fundamental component of human cognition, encompassing the encoding, organization, and retrieval of personally experienced events across the lifespan (Rubin, 2022). These memories are often centred around significant life events, which can be either *normative*—such as graduating from school or getting married—or *non-normative*, like experiencing a natural disaster or the COVID-19 pandemic (Habermas & Köber, 2015; Reitz, 2022). Non-normative events, particularly those that are unexpected and emotionally charged, tend to stand out in autobiographical narratives and exert a stronger psychological impact (Demiray & Bluck, 2020; Weststrate et al., 2022). Autobiographical memory is considered an integral component of human cognitive processes as it involves the encoding, organization, and retrieval of personally experienced events across the lifespan (Rubin, 2022).

In late 2019, the sudden emergence of a previously unanticipated and highly transmissible virus triggered a global health crisis. This novel virus-dubbed COVID-19 was quickly recognized as a quintessential Black Swan event (Taleb, 2007), defined by its unpredictability, severity of impact, and ability to be explained after the fact. By May 2023, the World Health Organization (WHO) had reported more than 770 million confirmed cases and more than 6.9 million deaths worldwide (WHO, 2024). On 11 March 2020, the WHO officially declared COVID-19 a pandemic, and by 8 September 2020, it had spread to 216 countries and exceeded 27 million confirmed cases (WHO, 2020). Although pandemics are not anomalies in the broader scheme of human history and have occurred for centuries (Marrazzo,

2025; Shchelkunov, 2009), from a psychological perspective they are deeply disruptive, non-normative life events. These events challenge established routines, personal expectations, and collective perceptions of security; often leaving profound psychosocial consequences.

Although the COVID-19 pandemic affected the entire globe, public health responses varied from country to country, depending on local policies and capacities. However, a common measure, was to restrict personal mobility—through lockdowns, curfews, and restrictions on public gatherings—to minimize contagion (Caglayan-Akay et al., 2025). Although these measures were effective in controlling viral transmission, they also disrupted daily routines and severely impacted psychological and emotional well-being (Aung et al., 2025; Enav et al., 2024). Thus, in the face of a global threat, governments relied on a common strategy that, while medically necessary, challenged the social and emotional stability to which people were accustomed.

In light of the aforementioned challenges, this chapter aims to analyze the *emotional states* and *psychological well-being* experienced during the COVID-19 pandemic, and to examine their impact on *cognitive processes* in order to facilitate a better understanding of autobiographical memory. To conduct this analysis, a narrative method will first be employed to explore and conceptualize the key psychological constructs involved. Subsequently, the PRISMA (Preferred Reporting Items for Systematic Reviews and Meta-Analyses) methodology will be applied to systematically review and synthesize the relevant scientific literature. The central research question guiding this review is as follows: *What is the evidence for the types of emotional states activated during and after the COVID-19 pandemic, and their impact on cognitive processes across different populations?*

1.1 Emotional State

In postmodern societies, emotions have gained central importance in both individual lives and broader social dynamics (Bauman, 2000; Giddens, 1991; Lyotard, 1979). Positive emotional experiences have been shown to play a central role in improving well-being, particularly in uncertain social contexts, as they foster adaptive responses, social connectedness, and long-term mental health (Tugade et al., 2021; Cregg & Cheavens, 2021; Seppälä et al., 2022). Therefore, given the critical role of emotional experiences in shaping human functioning, it is essential to define the construct (Gross, 2015; Keltner et al., 2019).

The scientific study of emotions in psychology has evolved through multiple paradigmatic shifts. In the late nineteenth century, Wundt (1897) conceptualized emotions as "affective compounds" structured along the dimensions of valence, arousal, and tension. However, early figures such as Titchener (1909) dismissed emotions because of their subjective nature, and James (1884), despite his theoretical contributions, failed to establish emotions as a core object of empirical study. Freud (1915) acknowledged the affective quality of emotions but did not center them in psychological research. The behaviorist movement further marginalized emotions, excluding them from scientific inquiry due to its focus on observable behavior. Only with the cognitive revolution did emotions regain prominence, reframed as empirically tractable processes involving neural, developmental, and functional dimensions. This shift reestablished affective phenomena as central to psychological science.

Contemporary approaches increasingly emphasize a constructive neuroscientific perspective, which views emotions not as fixed, biologically hardwired responses but as dynamic constructions shaped by brain–body processes, context, and prior experience (LeDoux, 2012, 2021; Barrett & Lida, 2024; Barrett, 2025). Indeed, we understand Emotion as a psychological process that interacts with affective, sensory, cognitive, and motor processes to produce diverse, unique and idiosyncratic emotional states. It will be exhausting to have emotional experiences all the time. Thus, emotional experiences occur only when a meaningful reality is perceived by the subject. The sensory processes—interoceptive and exteroceptive—transduce information about the world—internal and external—and, if it is meaningful, the affective processes manage the creation of arousal and valence effects. The body deals with the impact through inferential processes. These have the potential to generate responses in all structures. Depending on the type of emotional contextual category learned, the brain activates a particular emotional category by mental inference, creating a dynamic, specific and temporally limited conscious emotional state. And all these emotional experiences are more or less pleasurable (valence) and more or less arousing (arousal). In essence, emotions are not fixed responses; they are dynamic, context-dependent constructions that emerge from the interaction of brain, body, and environment. Emotional states arise when meaningful perceptions trigger affective and cognitive processes, shaping how we experience and adapt to the world.

1.2 Well-Being

Well-being is a multifaceted and evolving concept that refers to the state of optimal human functioning, encompassing the satisfaction of fundamental needs and the pursuit of personal growth and meaning.

Its roots can be traced back to classical philosophy, particularly in Aristotle's Nicomachean Ethics (Aristotle, 2009), where he distinguishes between hedonism—the pursuit of pleasure—and eudaimonism—the realization of human potential through the cultivation of virtue. This philosophical foundation aligns with Maslow's (1943) hierarchy of needs, which conceptualizes well-being as a progression from basic physiological and safety needs to higher level needs such as love and belonging, esteem, and ultimately self-actualization—the realization of one's full potential. Maslow argued that higher psychological goals can only be pursued once fundamental survival needs are met. Building on this hierarchical view, Carl Rogers (1961) proposed that well-being arises from the alignment between self-concept and lived experience. Similarly, Deci and Ryan's self-determination theory (2000) identified autonomy, competence, and relatedness as essential psychological needs that directly support the development of well-being. Likewise, Seligman's PERMA model (2011) introduced a multidimensional approach of well-being that includes positive emotions, engagement, relationships, meaning, and accomplishment. In a similar manner, Erikson (1960) framed well-being as the successful resolution of psychosocial tasks throughout the lifespan. Finally, Alderfer's ERG theory (1969) reformulated Maslow's structure into three dynamic categories—existence, relatedness, and growth—emphasizing that well-being is flexible and non-linear.

In summary, well-being is a dynamic and multidimensional construct that encompasses the fulfilment of basic needs and the pursuit of higher order values, shaped by personal, social, and developmental influences across the lifespan.

1.3 Cognition

Cognition refers to the collection of mental processes that allow humans to perceive, attend, remember, reason, solve problems, and use language. These processes enable individuals to acquire knowledge, make decisions, and adapt behaviorally to their environment (Browman, 2025; Kumar, 2025). Modern cognitive psychology considers cognition as a dynamic, distributed, and embodied system, rather than a purely internal or computational one. It involves constant interaction between the brain, body, and social context (Capurso, 2025). Furthermore, recent models emphasize that cognition is also shaped by developmental, emotional, and ecological factors, leading to more holistic and context-sensitive approaches to understanding human thought (Amodio, 2025; Zhang et al., 2025).

Cognition is best understood as a dynamic, embodied, and context-sensitive system shaped by emotion, development, and environment. Far from being isolated mental computation, it reflects the ongoing interaction between brain, body, and world. From a holistic perspective, the pandemic has engendered a novel lived reality, fundamentally shaped by the pervasive fear of contagion and the enforcement of mobility restrictions aimed at reducing transmission. This emergent context is rich in experiential meaning, eliciting diverse emotional responses that have significantly impacted individuals' well-being across both hedonic (pleasure-related) and eudaimonic (meaning- and purpose-driven) dimensions. These affective experiences are deeply interconnected with cognitive processes such as knowledge acquisition, decision-making, and psychological adaptation to uncertainty and change. Based on these considerations, the following section presents a PRISMA-based systematic review that was conducted to synthesize the relevant scientific literature.

2 Methods

2.1 Search Strategy

The literature search for various studies was conducted following the guidelines of the PRISMA model (Preferred Reporting Items for Systematic Reviews and Meta-Analyses) (Page et al., 2021) to ensure methodological transparency, rigor, and replicability. The final search was carried out in March 2025. The electronic databases included were as follows: Picador, Psychology and Behavioral Sciences Collection, PsycINFO, PubMed, Scopus, and Web of Science. The search terms used were: "covid" OR "covid-19" OR "sars-cov-2" OR "sars" AND "emotion" OR "emotions" OR "emotional state" AND "psychological well-being" OR "well-being" AND "cognitive processes" OR "cognition". To ensure methodological rigor, two reviewers independently screened the studies and extracted data. Disagreements between the reviewers were resolved by discussion and consensus.

2.2 Study Selection and Eligibility Criteria

The initial selection of publications was based on the title, abstract, and keywords. Eligible articles were identified according to specific inclusion and exclusion criteria. Inclusion criteria included studies that addressed emotional state and well-being during the pandemic and their impact on cognition, and articles published in English. Consequently, the exclusion criteria included publications in languages other than English, articles with restricted access, studies not related to emotional state, well-being, or cognitive processes, and studies that were done during COVID but do not analyze the effect of the pandemic (Figure 1).

3 Results

3.1 Prisma

To understand the impact of the COVID-19 pandemic on emotional states and well-being, and how these factors have influenced cognitive functioning, we conducted a systematic review according to PRISMA. From the initial pool of studies, 43 studies met inclusion criteria and were retained for qualitative synthesis. The included studies varied widely in terms of geographical origin, participant characteristics (adults, adolescents, students, older adults), study design (cross-sectional, longitudinal, descriptive, qualitative), and the psychological constructs assessed. An overview of the characteristics and key findings of each study is presented in Table 1. Below, we provide a narrative synthesis of the findings, structured around three main thematic

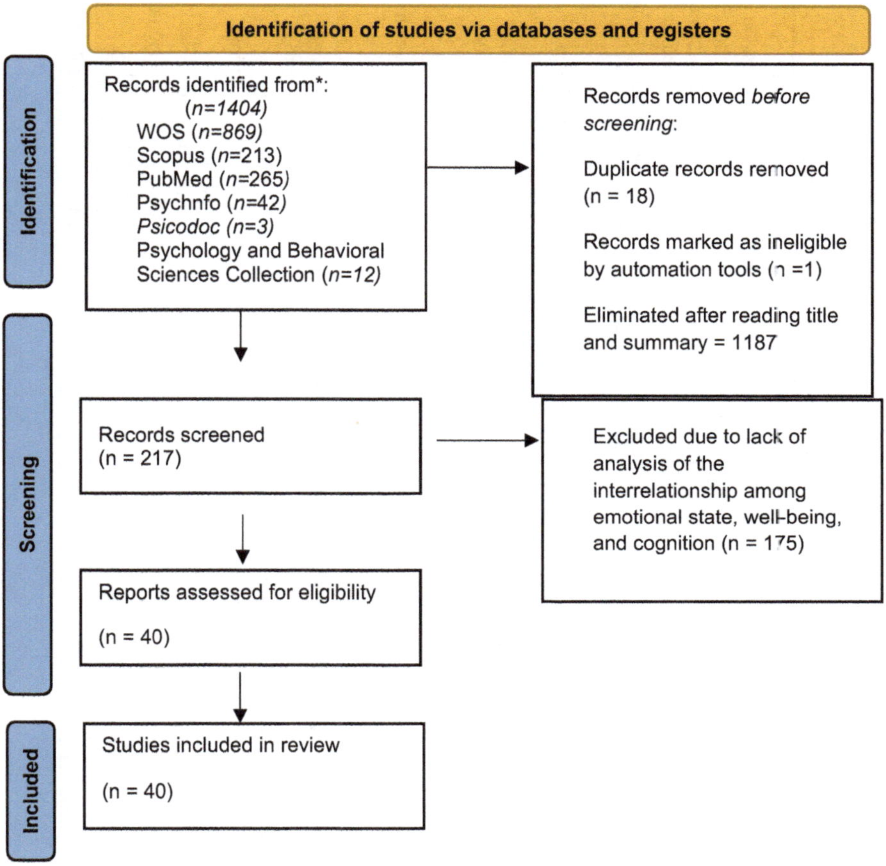

Fig. 1 PRISMA flow-chart of the present review

domains: (1) affectivity dimensions and emotional states, (2) cognitive functioning and appraisal, and (3) psychological well-being.

3.2 *Affectivity Dimensions and Emotional States*

The emotional impact of the pandemic was widely documented across populations. Overall, findings indicate that the pandemic initially triggered heightened negative emotional states, which in many populations gradually shifted as individuals adapted to the evolving crisis. Longitudinal studies provided compelling evidence of this dynamic. Dumciene and Pozeriene (2022), who compared emotional frequency across the first and second waves in Lithuania, showed a relative increase in positive emotions and decrease in negative ones during the second wave. A similar trend was observed by Kyröläinen and Kuperman (2024).

Table 1 Descriptions of the studies included in the present review

References	Data collection time	Design	Sample	Objective	Assessment tool	Outcome
Low et al. (2021)	**New Zealand.** Study 1= June 2018–June 2019 Study 2= between April 8 and 27, 2020	Longitudinal	Study 1: 468 parents (M age = 37.1). Study 2: 1607 adults (365 from Study 1; remaining M age = 51.6; 65.2% female; 71.6% parents).	To assess how emotion regulation strategies during COVID-19 related to psychological and physical health, compared to pre-pandemic levels.	Study 1: CES-D, RAND-36, and ad hoc measures on well-being, fatigue, health, stress, and emotion regulation. Study 2: NZAVS-based; Kessler-6, SF-36, and similar measures of emotion regulation and well-being.	Rumination and emotional suppression were linked to higher depression and distress, and lower well-being. Rumination also predicted greater fatigue. Cognitive reappraisal showed mixed results, benefiting well-being only in one study. No overall change in distress was found between pre-lockdown and lockdown.
Vanaken et al. (2022)	**Belgium.** T1= February 2018; T2= March 2020	Longitudinal	N1= 278 (88.3% female, 17–22, M age = 18.46, SD=0.78). N2=198 (174 female, M age = 20.61).	To examine narrative coherence as a general cognitive skill linked to resilience beyond specific traumatic events.	Online writing task; Flourishing Scale (FS); Social Support List–Interactions (SSL-I).	Emotional well-being and support declined. Narrative coherence predicted higher well-being. Emotional loneliness rose—mainly in older adults—due to personal loss and pandemic concerns. Mental health stayed stable.

Van Tilburg et al. (2021)	**Netherland.** May 2020.	Longitudinal	1876 (65 or older, M age = 73, 51% men).	To examine how reduced social contact, personal loss, and perceived societal threats impacted well-being.	Ad hoc questionnaires for Social and Emotional loneliness, Contact frequency, and Active Coping. Mental Health Inventory (MHI-5, Berwick et al., 1991).	After 7 months, mental health remained stable. Emotional loneliness rose moderately—especially in older adults—driven by loss, pandemic concerns, and low institutional trust, not social distancing.
Kivi et al. (2021)	**Sweden.** 2015, 2016, 2017, 2018, 2019. The 2020 wave, collected between March 26 April 2.	Longitudinal	1071 (adults, M age = 36.54. 47.34% women).	To investigate early pandemic effects on worry, risk perception, social distancing, and their longitudinal impact on well-being.	Life satisfaction. Financial satisfaction. Self-rated health. Loneliness. Health worry. Societal worry. Financial worry. Societal risk. Risk of being infected. Social distancing.	Most participants worried about societal (69.5%), health (44.9%), and financial (25.1%) impacts; 86.4% perceived high societal risk and 71.2% practiced social distancing. Well-being remained stable or improved (health, finances). Health/ financial worry predicted lower well-being; societal worry and distancing predicted higher well-being.

(continued)

Table 1 (continued)

References	Data collection time	Design	Sample	Objective	Assessment tool	Outcome
Fernández-Abascal & Martín-Díaz (2022)	**Spain.** During March-April 2020.	Longitudinal (online)	303 (M age = 36.54. 19.8% men, 80.2% women).	To assess changes in hedonic and eudaimonic well-being during compulsory COVID-19 confinement.	*Well-being*: PERMA-Profiler (Butler & Kern, 2016). Positive Emotion, Engagement, Relationships, Meaning, and Accomplishment. *Dispositional optimism*: The Life Orientation Test-Revised (Scheier et al., 1994), designed to assess generalized expectations of positive and negative outcomes.	Well-being scores (positive emotion, engagement, meaning, overall) were higher before lockdown (M1); negative emotion increased during lockdown (M2).
Peng and Roth (2022)	**USA.** 2016 (1st) and June of 2020 (2nd).	Longitudinal.	1141 (older adults M age = 67, 47% men).	To assess changes in social, physical, digital isolation, and loneliness in U.S. adults aged 50+ before and during lockdown.	Loneliness (Revised UCLA Loneliness Scale, Russell, 1996). Instrumental functional limitation (IADL scale, Lawton & Brody, 1969).	Physical and social isolation increased, especially among those with high COVID-19 concern. Digital isolation and loneliness remained unchanged.

Kyröläinen & Kuperman (2024)	**USA and Canada.** Eight sessions from March 2019 with intervals of 2–5 months.	Longitudinal psycholinguistic.	1311 (adults, M age = 62.51; female = 911).	To examine emotional well-being and distress through linguistic markers in autobiographical narratives across pre-, during, and post-lockdown periods.	Written individual narratives and analyzing with the data of the Cognitive and Social Well-Being (CoSoWELL) corpus.	Positive emotions declined and negative emotions rose, peaking 7 months post-lockdown, then returning to baseline. Older adults used more COVID-related and fewer loneliness-related terms.
Castillo et al. (2024)	**USA.** T1; May 3, 2020 to Nov. T2; Aug 13, 2021 to Oct 7, 2021)	Longitudinal	T1= 1810 (M age = 32.02, 851 Female). T2 = 726 (M age = 34.77, 345 Female).	To assess the accuracy and consistency of emotional memories and how event type influences emotional recall reliability.	Ad hoc COVID-related stressors questionnaire. Perceived Stress Scale (PSS; Cohen et al., 1983) and 21-item short-form Depression Anxiety Stress Scale (DASS-21; Henry & Crawford, 2005).	One-year post-outbreak, participants reported less negative emotion but overestimated past emotional intensity. This memory bias was influenced by current emotional state and well-being.
Matic et al. (2022)	**Slovenia.** First wave= 23 July to 31 December 2020. Second wave= 1 January to 23 February 2022	Longitudinal (Online)	First wave n= 1785 (77.0% female). Second wave= 1241 (78.3% female). Age= 18–32	To compare depression, anxiety, stress, and suicidal ideation in young adults between the early pandemic and the Omicron wave.	Depression, Anxiety, and Stress Scale—21 Items (DASS-21); The Suicidal Ideation Attributes Scale (SIDAS)	During the Omicron wave, depression, anxiety, stress, and suicidal ideation were significantly higher than in early pandemic stages, across all groups. Social stressors, not infection, drove distress.

(continued)

Table 1 (continued)

References	Data collection time	Design	Sample	Objective	Assessment tool	Outcome
Zaninotto et al. (2022)	**Italy and United Kingdom.** 4 weeks during the initial stage of lockdown in 2020	Longitudinal	Italy (males n=32, age= 40±14; females n=39, age= 41±15). United Kingdom males= 5, age 43±17; females n=20, age 23±6) 4 weeks during the initial stage of lockdown in 2020	To explore how internal traits and external factors affect well-being and cognitive functioning during the pandemic.	Big Five, IES-R, FFMQ, SCL-90-R (short), DASS-21, and ISI, along with questions on habits and substance use. Two cognitive tasks assessing attention and vigilance were conducted via PsyToolkit.	Initially, well-being was shaped by internal factors; later, external behaviors became more influential. Positive factors included TV, internet use, dog walking, and outdoor access; negative ones were social media and video calls. No impact on cognitive performance was found.
Perica et al. (2021)	**USA.** 2019, 2020	Neuroimaging-two multimodal accelerated longitudinal studies	Post-covid group: n = 286 (age 10–31; 152 female). Pre covid: n= 111 (age=10–31). Two age groups: adolescents (10–18.6) and young adults (18.6–31)	To analyze how stress-related brain connectivity predicts real-life stress during the pandemic, comparing adolescents and young adults.	MR data and ABCD COVID-19 (Worry score; Sadness score; COVID stress score; Positive affect)	Worry and COVID-19-related stress rose with age. In individuals under 18, stronger connectivity between the vmPFC and hippocampus was linked to higher stress and worry.

Sikka et al. (2024)	**United Kingdom and Australia.** July-November 2020 (2-week report period).	Longitudinal	153 (M age = 48.42, 23 men)	To examine how COVID-19-related concern, anxiety, and worry relate to affect during mind-wandering and dreaming, controlling for sleep quality.	Daily logs for 2 weeks tracking mind-wandering, dreams, nightmares, COVID-19-related concerns, anxiety, and sleep quality, using Likert scales and standardized questionnaires (General COVID-19-Related Concern, WBQ; Anxiety-Coronavirus anxiety scale, WBQ; General COVID-19-Related worried)	COVID-19 concern and anxiety did not affect mind-wandering, but daily COVID-19 worry was linked to more negative and less positive affect in both mind-wandering and dreaming. Sleep quality predicted dream affect. Younger people and those seeing themselves at risk reported more negative affect; those with at-risk close ones showed less.
Dumciene and Pozeriene (2022)	**Lithuania.** First survey from 20 June 2020 until 15 July 2020. Second survey from 1 July 2021 to 15 July 2021	Longitudinal	Master students *First wave n=118 (67 females and 51 males) *second wave n=128 (74 females and 54 males)	To compare changes in emotions, coping strategies, and psychological well-being between the first and second COVID-19 waves.	Scale for Measuring Psychological Well-Being. Multidimensional Emotion Questionnaire. Coping Measure Scale (COPE-14)	Negative emotions were stronger during the first wave. Coping strategies were mostly problem- and emotion-focused. After the second wave, positive emotions increased, negative emotions decreased, and well-being significantly improved.

(continued)

Table 1 (continued)

References	Data collection time	Design	Sample	Objective	Assessment tool	Outcome
Jiang (2020)	**China.** During the peak of the COVID-19 outbreak in China.	Longitudinal (online)	231 adults. Mean age 44.74 years. 69% female.	To examine the benefits of feeling gratitude every day during the COVID-19 outbreak.	Feeling gratitude (likert). Positive and negative affect (short version of the Affect Valuation Index, Tsai et al., 2006). Perceived stress related to COVID-19 (adapted Perceived Stress Scale, Cohen et al., 1983). Life satisfaction (Satisfaction with Life Scale, Diener et al., 1985).	Daily gratitude was linked to more positive affect, lower COVID-related stress, and better health on the same and next day. Over 14 days, it predicted more positive and less negative affect, but not long-term life satisfaction or health.
Ang et al. (2022)	**New York (USA).** First semester of 2020	Cross-sectional (Online)	476 private university first year	To examine how hope and gratitude help mitigate the negative mental health effects of COVID-19.	Adult Hope Scale (HS). Gratitude Questionnaire (GQ-6). Ad hoc COVID-19 protective self-efficacy. Coronavirus Experiences and Impacts Questionnaire (CEQ). Satisfaction with Life Scale (SWLS). Depression and Anxiety Scales-21 (DASS-21).	Greater negative psychological impact from COVID-19 was linked to poorer mental health, whereas hope and gratitude were associated with reduced negative impact.

Folville et al. (2023)	**Belgium.** In April 2020. Pre-lockdown period March 2019).	Cross-sectional (Online)	188 (M age = 42.3, 144 women).	To explore how lockdown-related changes in well-being relate to experiential diversity and autobiographical memory.	(1) demographic and COVID-19-related information, and of affective states, anxiety, psychological well-being, and distress; (2) questions on sleep quality, fatigue, and mental workload; and (3) cue-word-based recall of autobiographical memories from before and during the lockdown.	Lower experiential diversity during lockdown was linked to reduced well-being. While it didn't affect memory richness or intensity, well-being correlated with the subjective richness of memories, which shifted from work to home themes.
Padmanabhanunni and Pretorius (2024)	**South Africa.** Undefined 2022.	Cross-sectional (Online)	322 (students, M age = 26, 77% women).	To examine how problem-solving appraisal influences fear of COVID-19, PTSD symptoms, and their link with anxiety.	Fear of COVID-19 Scale (FCV-19S); Problem-Solving Inventory (PSI); The PTSD Checklist for DSM-5 (PCL-5); and the five-item short version of the trait scale of the State–Trait Anxiety Inventory (Stai-T5)	Fear of COVID-19 was linked to poor problem-solving self-appraisals, higher PTSD, and anxiety. Viewing oneself as an effective problem solver reduced both fear and PTSD. Anxiety moderated this relationship.

(continued)

Table 1 (continued)

References	Data collection time	Design	Sample	Objective	Assessment tool	Outcome
Nitschke et al. (2021)	**Austria.** In April 2020.	Cross-sectional (Online)	902 (M age = 49.60, 55.6% women).	To examine how stress, general and COVID-specific worry, and fatigue relate to social connectedness during lockdown.	Social Network Index (SNI);Perceived Stress Scale (PSS);Penn State Worry Questionnaire (PSWQ); Chalder Fatigue Questionnaire (CFQ);COVID-19-specific worries	Greater social connectedness was linked to lower perceived stress and fewer COVID-specific worries, helping to reduce distress.
Mato et al. (2021)	**Worldwide.** Between mid-April 2020 and mid-May 2020	Cross-sectional (Online)	4057 (21 countries, M age=41.45, 80.8% of female).	To investigate how social connectedness and disconnection affect the link between perceived COVID-19 threat and psychological outcomes like post-traumatic growth and stress.	Compassionate Engagement and Action Scales and the Social Safeness and Pleasure Scale; Fears of Compassion Scales and the UCLA Loneliness Scale; The Perceived Coronavirus Risk Scale (PCRS); Post-traumatic Growth Inventory (PGI);Impact of Event Scale—Revised (IES-R)	Perceived COVID-19 threat was linked to both post-traumatic growth and stress. Social connection amplified growth and stress, while disconnection increased stress and reduced growth. These effects were consistent across countries.

Naaz et al. (2022)	**Saudi Arabia.** Academic year 2020–2021—Fall semester	Cross-sectional (Online)	261 (undergraduate medical students, M age= 19.63, 54% female).	To identify students' emotional responses and coping strategies, and examine their correlations.	Partially validated questionnaire with three sections: 1) demographics 2) ad hoc questionnaire about emotional reactions (scared, afraid, angry and sad) during the lockdown COVID-19, 3) short COPE (Coping Oriented to Problems Experiencing)	Anxiety and sadness were highest in fifth-year students; fear in first and fourth years; anger in first years. Avoidant coping was linked to higher levels of all emotions and predicted anger and sadness.
Diotaiuti et al. (2023)	**Central and South of Italy.** Between April and May 2020.	Cross-sectional (Online)	544 (M age = 31.59 71.4% female).	To examine how lockdown-related concerns shape cognitive appraisals, influence self-efficacy, and relate to psychological distress.	Ad hoc Likert-scale questionnaires were used to assess internal concerns during lockdown, external concerns, individual cognitive appraisal patterns (threat and coping appraisals), perceived psychological distress, and generalized self-efficacy (GSE).	Distress was higher in women, chronically ill individuals, and those with vulnerable contacts. It was reduced by self-efficacy and adaptive appraisals, but increased by maladaptive threat appraisals. Younger age predicted more distress.

(continued)

Table 1 (continued)

References	Data collection time	Design	Sample	Objective	Assessment tool	Outcome
Kodami et al. (2021)	**Africa, North America, Asia, Australia, Europe, South America.** From February 2020, until July 2020	Cross-sectional (Online)	2185 (Africa, n=554, 266 female; North America n = 497, 244 female; Asia n= 881, 425 female; Australia n = 494, 270 female; Europe n = 527, 270 female; South America n = 119, 65 females).	To examine how COVID-19-related declines in quality of life are linked to perceived stress and emotion regulation.	Demographic Questionnaire; Covid-19 QoL questionnaire; Emotion Regulation questionnaire; Perceived stress questionnaire	Quality of life declined over time, while stress and emotion regulation difficulties increased—especially among younger individuals and those with greater declines in well-being. Longer quarantine was expected to worsen these effects.
Xu and Chen (2024)	**China.** Between April and May 2023	Cross-sectional (Online)	986 (Age = 18 or older, 51.7% female).	To examine how coping strategies moderate the link between infodemic-driven healthcare overuse and psychological outcomes like depression and cyberchondria.	Cyberchondria Severity Scale-12; Generalized Anxiety Disorder Assessment; Coping Orientation to Problems Experienced Inventory; Infodemic- and Misinformation-Driven Overuse of Healthcare Services	Overusing healthcare services during the pandemic was linked to higher levels of depression and cyberchondria.

Valiente et al. (2021)	**Spain.** In April 2020.	Cross-sectional (Online)	1951 (M age= 45.16, 918 females).	To determine how the lockdown affected well-being and negative mental health outcomes.	An ad hoc questionnaire assessed COVID-19-related anxiety, economic threat, health status, substance use. Three-Item Loneliness Scale, TILS, PHQ-9, GAD-7, and the Pemberton Happiness Index.	22.1% showed probable depression and 19.6% clinical anxiety. Older adults had lower distress. Income was the strongest predictor of well-being; lower well-being was linked to female gender, younger age, prior mental health issues, and COVID-related anxiety.
Yan et al. (2022)	**China.** May 17–June 18 2022	Cross-sectional (Online)	402 (university students, M age=22.04, 238 females).	To investigate how anxiety influences academic involution among university students during COVID-19.	COVID-19 perceived risk scale, perceived stress scale (PSS-10), generalized anxiety disorder scale (GAD-7), brief need for closure scale (NFCS-15), and academic involvement behavior scale (AIB-15)	Anxiety and stress from perceived COVID-19 risk were linked to academic involution, mediated by students' need for cognitive closure.

(continued)

Table 1 (continued)

References	Data collection time	Design	Sample	Objective	Assessment tool	Outcome
Yang et al. (2020)	19–21 February 2020	Cross-sectional (online)	3159 (older 18, 51% female)	To examine how social media use and emotion regulation strategies relate to mental health during the COVID-19 pandemic.	Multiple Happiness Questionnaire (MHQ); General Health Questionnaire (GHQ-20); self-designed items for social media activities; Emotion-regulation questionnaires (Liu and Liu)	Accurate, positive COVID-19 content, limited discussion, and empathy improved well-being. Social media's impact depended on emotion regulation, with cognitive reappraisal buffering negative effects.
Bland et al. (2022)	**United Kingdom (Manchester).** Between April and May 2020.	Cross-sectional (online)	92 (adults, 50 females).	To assess how COVID-19 social isolation affected emotional and social cognition, especially in those with greater disruption to usual social connections.	Ad hoc COVID-19 social isolation questionnaire. EMOTICOM neuropsychological test battery (emotional recognition task, affective go no-go task, cooperative behavior task).	Social isolation impaired affective cognition, reducing positive bias in emotion recognition and attention to emotional faces—especially with less contact with friends. Frequent contact increased cooperative behavior.

Hou et al. (2021)	**China.** In February 2020.	Cross-sectional (online)	5115 (from the general public across 32 provinces and autonomous regions in China).	To investigate how quarantine length relates to negative affect, with worry and anticipation as mediators, and optimism, uncertainty tolerance, social support, and healthy behavior as moderators.	Depression and Anxiety subscales of Brief Symptom Inventory (Derogatis, 2000). Three items adapted from the McCaul Brief Worry Scale (McCaul and Goetz, n.d.). Dispositional optimism (Life Orientation Test-Revised. Scheier et al., 1994). Intolerance of Uncertainty–Short Form (Carleton et al., 2007). Social support (five items adapted from the Medical Outcomes Study Social Support Survey, Sherbourne and Stewart, 1991). Ad hoc ítems for anticipation of contagion and healthy behavior.	Worry and anticipation mediated the link between quarantine length and distress. Optimism, uncertainty tolerance, and social support buffered the effects of quarantine on depression and anxiety.

(continued)

Table 1 (continued)

References	Data collection time	Design	Sample	Objective	Assessment tool	Outcome
Perasso et al. (2023)	**Italy.** In March 2022.	Cross-sectional (online)	861 (adolescents, 12–19 years, Female 58%).	To analyze the emotions experienced by Italian adolescents during the pandemic and the types of social support perceived, using quantitative and qualitative methods.	Ad hoc questionnaire. The survey presented four questions, in the following areas: i.e., most frequent negative emotions during the pandemic (e.g., "Which emotions did you feel more often during the COVID-19 pandemic?"	Adolescents reported boredom, sadness, fear, and agitation, but found support in family, friends, and themselves. Positive emotions arose from self-awareness, personal growth, and relational skills.
Klaiber et al. (2021)	**USA.** From March 18 to April 11, 2020	Cross-sectional	776 (330 younger 19–39 years, 253 middle-aged, 40–59 years) and older, 60–91 years, Female 86.4%).	To examine age differences in exposure and emotional reactivity to daily stressors and positive events during the early weeks of the COVID-19 outbreak.	COVID-19 threat (stress appraisals regarding the threat of COVID-19; Pow et al., 2016, 2017; adapted from Folkman et al., 1986). Positive and negative daily affect (PANAS-X; Watson & Clark, 1999). Daily stressors (modified version of the Daily Inventory of Stressful Events; Almeida et al., 2002). Daily and past week positive events (ad hoc).	Younger adults had more COVID-related concerns, less positive affect, and higher emotional reactivity than older adults, despite similar exposure. They benefited more from positive events, while older adults remained more emotionally stable.

Fingerman et al. (2021)	**USA.** During May and June 2020.	Cross-sectional	226 (older adults. M age = 77.2, Female 55.3%).	To examine how living alone affected daily social contact and emotional well-being in older adults during the pandemic.	Demographics. COVID experiences: symptoms. Daily experiences (Day Reconstruction Method DRM; Kahneman et al., 2004; Oerlemans et al., 2011). Social engagement (ad hoc). Emotion (ad hoc).	Older adults living alone had less in-person contact and help exchange, but felt more positive during in-person interactions. Phone contact increased negative affect only in those living alone, who more often connected with friends than family.
Dattilo et al. (2022)	**USA.** From August to December of 2020	Cross-sectional	225 (college students, M age = 18.95, Female 75.6%).	To assess whether perceived COVID-19 impact moderated the link between dating anxiety and depressive symptoms.	Demographics. Dating anxiety (Dating Anxiety Scale for Adolescents DAS-A; Glickman & La Greca, 2004). Perceived impact of COVID-19 (COVID-19 Exposure and Family Impact Survey—Impact Subscale for Adolescents and Young Adults CEFIS; Kazak et al., 2020). Depressive symptoms (Center for Epidemiological Studies—Depression Scale CES-D; Rodolff, 1977).	Dating anxiety was associated with higher depressive symptoms, especially when the perceived impact of COVID-19 was greater.

(continued)

Table 1 (continued)

References	Data collection time	Design	Sample	Objective	Assessment tool	Outcome
Barber and Kim (2021)	**USA.** During March 2020.	Cross-sectional	146 (younger, 18–35 years, 68 men), 156 (older adults, 65–81 years, 84 men).	To test whether older adults–and particularly older men—would report the least amount of COVID-19 worry and also fewer COVID-19 behavior changes.	COVID-19 risk perception, worries and behavior changes (ad hoc).	Despite perceiving higher COVID-19 risk, older men in the U.S. were less worried and made fewer behavioral changes than other groups.
Robbins et al. (2022)	**USA.** Between June and October 2020.	Cross-sectional	3122 (older adults, aged 65 and older, Female 54.5%).	To assess behavioral changes, such as changes to sleeping or eating due to the COVID-19 pandemic, and their implications for emotional well-being.	Ad hoc. Changes to daily behaviors attributed to the COVID-19 pandemic (i.e., sleep quality, sleep duration, watching television, walking as exercise, doing vigorous activities, consuming alcohol, eating, and smoking or vaping). subjective feelings of distress about the COVID-19 pandemic (participants were asked to report feelings of depression and anxiety about COVID-19.	Among older adults, 22.7% reported depression and 28.3% anxiety related to COVID-19. Women reported higher levels of both. Poor sleep, irregular sleep, more TV, and less walking were linked to greater distress.

Bhargav and Swords (2022)	**Ireland.** Between February and March 2021.	Cross-sectional (online)	321 (college students, aged 18–25, 78.8% female).	To assess COVID-19-related stress in college students and identify sociodemographic and psychosocial factors linked to greater distress.	The COVID-19 Student Stress Questionnaire (Zurlo et al., 2020). Psychological distress (Clinical Outcomes in Routine Evaluation—CORE-10; Barkham et al., 2015). Adverse childhood experiences (adapted version of the Early Adverse Experiences Questionnaire (Felitti et al., 1998). Resilience (Brief Resilience Scale (BRS) developed by Smith et al. (2008).	Female students, those with chronic illness, financial concerns, or prior psychological issues reported greater COVID-19-related stress, indicating the need for targeted support in higher education.
Khodami et al. (2022)	**Worldwide.** During February 20, 2020, until July 20, 2020.	Cross-sectional (online)	3002 (50.2% women).	To investigate and predict changes in quality of life and psychological outcomes worldwide during the pandemic.	The Covid-19 QoL [13] is a six-item scale covering primary regions of Quality of life comparable to emotional wellness. Emotion Regulation questionnaire (reappraisal and suppression). Perceived Stress Scale (PSS-10).	Quality of life declined and perceived stress increased over time, with greater emotion regulation difficulties. Younger individuals and those with reduced quality of life experienced more stress. Longer quarantine predicted further declines in well-being and increased stress.

(continued)

Table 1 (continued)

References	Data collection time	Design	Sample	Objective	Assessment tool	Outcome
Dooley et al. (2024)	**Ireland.** Between September and December 2021.	Cross-sectional	4404 (adolescents, age <16, 51.2% men).	To assess the prevalence of mental health problems, demographic variations, and the impact of COVID-19 lockdowns on mental health.	General mental health, emotional and behavioral problems (Strengths and Difficulties Questionnaire, SDQ) Depressive, anxiety (Symptom Check List 90) and psychotic (Psychotic-like experiences, PLEs) symptoms. Self-reported rates of self-harm, suicidal ideation, and suicidal attempt (ad hoc).	Over a quarter of adolescents rated their mental health as poor, with high rates of self-harm (39%), suicidal ideation (42%), and suicide attempts (11%). Gender-diverse youth and females showed worse mental health outcomes than cisgender males.

Cross-sectional studies converged with these findings, although offering only snapshots of emotional functioning. In an international sample of over 2000 participants, Khodami et al. (2022) found elevated levels of perceived stress and emotion-regulation difficulties, particularly among younger individuals and those experiencing reduced quality of life. Naaz et al. (2022), focusing on Saudi medical students, showed that negative emotions such as fear, sadness, and anger varied across academic years, with anger most prevalent among first-year students and fear more pronounced in early and mid-level cohorts. The study also found that avoidant coping strategies—more frequently used by female participants—were strongly associated with heightened emotional distress. Additional evidence for the role of coping strategies comes from Hou et al. (2021), who explored how the length of quarantine predicted depression and anxiety through cognitive mediators such as worry and anticipation. Notably, the negative effects of prolonged isolation were buffered by dispositional optimism, tolerance of uncertainty, social support, and engagement in health-promoting behaviors.

Worry and stress-related emotional states were particularly salient among student populations. Yan et al. (2022) found that Chinese university students reporting higher perceived COVID-19 risk also experienced elevated anxiety and stress, which contributed to a pattern of academic involution—a maladaptive state characterized by excessive effort in the absence of external rewards. Similarly, Perica et al. (2021), using a longitudinal neuroimaging design, demonstrated that increased COVID-related stress and worry were associated with stronger functional connectivity between the ventromedial prefrontal cortex and the posterior hippocampus in youth, suggesting a neurobiological basis for affective dysregulation under stress.

Gratitude and hope emerged consistently as protective emotional dispositions. In a daily diary study, Jiang (2020) found that higher reported levels of gratitude were associated with increased positive affect and reduced negative affect across 14 days, although this benefit did not extend to life satisfaction or health at the 1-month follow-up. Ang et al. (2022) replicated and extended these findings by showing that both gratitude and hope predicted lower levels of COVID-19-related psychological impact in a large cross-sectional sample.

Hope and emotional growth were also observed among adolescents. Perasso et al. (2023), using a mixed-methods design, found that Italian adolescents reported experiencing boredom, fear, and sadness, but also expressed signs of personal development such as self-awareness, relational growth, and increased emotional insight.

Another affective dimension that received attention was social connectedness. Nitschke et al. (2021), analyzing data from Austria, reported that higher levels of social connection were associated with lower perceived stress and fewer COVID-specific worries. Similarly, Mato et al. (2021), in a large international sample spanning 21 countries, found that perceived COVID-19 threat was linked to both increased stress and post-traumatic growth. Notably, social connection amplified both responses, while social disconnection was associated exclusively with increased stress.

Age-related differences in affective responses were a common finding. Klaiber et al. (2021) found that younger adults reported more negative affect, less positive affect, and greater emotional reactivity to both COVID and non-COVID stressors, compared to older adults. While older individuals showed lower affective reactivity, the emotional benefits of positive events were more pronounced in younger adults. Fingerman et al. (2021), focusing on older adults in the United States, found that those living alone experienced greater negative emotions during phone-based social contact compared to those engaging in face-to-face interaction.

Finally, individual differences in emotional response were also reported. Bhargav and Swords (2022), studying Irish university students, found that female gender, chronic illness, and financial concerns were significant predictors of COVID-related stress and negative affect. Dattilo et al. (2022) noted that dating anxiety and gender influenced the perception of COVID-related threat and behavioral change among college students in the U.S. Barber and Kim (2021) showed that although older adults in the U.S. perceived higher risk, older men reported lower worry and made fewer behavioral changes than other groups, pointing to age–gender interaction effects in emotional responsiveness.

3.3 Autobiographical Memory and Cognitive Processes

Several studies included in this review explicitly addressed how autobiographical memory and narrative processes were influenced by, and in turn influenced, emotional well-being. Vanaken et al. (2022) showed that greater narrative coherence in personal stories was linked to higher emotional well-being, supporting the view that structured meaning-making can buffer against distress. Kyröläinen and Kuperman (2024), using a psycholinguistic approach to analyze over 1300 autobiographical narratives from older adults in the U.S. and Canada, documented an inverse-U shaped pattern in emotional valence. Negative affect—including sadness, disgust, anger, and fear—increased sharply during the early lockdown phase and later decreased, while positive emotions showed the opposite pattern. A marked drop in optimism and rise in negative affect during the first year of lockdown mirrored broader public trends of growing pessimism and anxiety.

The importance of experiential variety and meaning-making was demonstrated by Folville et al. (2023), who showed that reduced engagement in diverse and stimulating activities during lockdown was linked to lower well-being. Interestingly, while experiential richness did not predict the intensity of autobiographical memory, it was strongly associated with its subjective richness, which in turn correlated with well-being.

Memory distortions were another relevant cognitive outcome. Castillo et al. (2024) compared emotional memory related to the 9/11 attacks and the COVID-19 pandemic. Their study revealed that 1 year after the onset of the pandemic, participants tended to overestimate the intensity of their early emotional experiences, a pattern not observed for 9/11 memories.

The pandemic profoundly affected cognitive functioning across multiple domains, including appraisal, attention, problem-solving, memory, emotion regulation, and spontaneous thought. A major thematic focus across studies was the role of cognitive appraisal in shaping emotional and behavioral outcomes. Padmanabhanunni and Pretorius (2024) conducted a study among South African university students and found that fear of COVID-19 was significantly associated with poor self-appraisals of problem-solving ability, which in turn predicted elevated levels of anxiety and PTSD symptoms. Notably, anxiety also moderated the relationship between fear and problem-solving appraisal, suggesting a feedback loop between emotional arousal and perceived cognitive capacity. Diotaiuti et al. (2023) reinforced this perspective, demonstrating that certain appraisals—such as perceiving lockdown as a threat—were associated with greater psychological distress; in contrast to functional appraisals of the situation, such as viewing it as a challenge, enhanced self-efficacy, and reduced distress. These results point to the importance of interpretative frameworks in modulating the psychological consequences of stress exposure.

The role of emotion regulation, particularly as a cognitive-affective strategy, was a key element in several studies. Low et al. (2021), in a longitudinal study conducted in New Zealand, assessed pre-pandemic and intra-pandemic changes in health and regulation strategies among family groups. They found that while general health and emotion regulation remained relatively stable, participants reported an increase in the use of cognitive reappraisal during lockdown, along with decreased fatigue. These changes suggest that individuals may engage in adaptive cognitive reframing to mitigate the impact of stress under lockdown conditions. Vanaken et al. (2022) further elaborated on this notion by showing that individuals with more coherent personal narratives—reflecting a more structured and integrated cognitive–emotional processing style—displayed greater emotional stability during the pandemic.

A different line of research focused on spontaneous thought processes, such as mind-wandering and dreaming. Sikka et al. (2024), in a longitudinal diary study, explored how COVID-related worry influenced affect during mind-wandering and sleep. Their findings revealed that daily COVID-19 worry predicted more negative and less positive affect during both waking and dreaming states. Interestingly, participants who perceived themselves at high risk for infection showed more negative affect during these spontaneous thought episodes, while those concerned about close others at risk showed less negative affect.

Information processing and infodemic overload were also highlighted as significant stressors with cognitive consequences. Huan and Chen (2024) found that individuals who engaged in excessive and unregulated healthcare use due to COVID-19 misinformation (a phenomenon labeled as "infodemic-induced health misuse") showed higher levels of cyberchondria and depressive symptoms.

3.4 Well-Being

The COVID-19 pandemic has posed a profound challenge to individual and collective well-being. However, empirical literature reveals mixed results. While some populations and timepoints exhibited marked declines in subjective well-being, others maintained relative stability or even reported gains.

Several longitudinal studies found evidence of declining well-being, particularly during the initial lockdown periods. Fernández-Abascal and Martín-Díaz (2022), in a Spanish sample, reported a significant decrease in multiple domains of well-being, including positive emotion, engagement, and meaning, during lockdown compared to pre-pandemic levels. Simultaneously, scores for negative emotion increased, indicating both a reduction in flourishing and a rise in distress. Similarly, Van Tilburg (2021) conducted a large-scale longitudinal study in the Netherlands with adults aged 65 and older, finding that although general mental health remained stable, social and emotional loneliness increased significantly over 7 months. Emotional loneliness was closely associated with personal loss, fear of infection, and declining trust in institutions.

In a longitudinal comparison across pandemic waves, Matić et al. (2022) reported a substantial increase in depression, anxiety, stress, and suicidal ideation during the Omicron wave. This pattern held across gender, age, and lifestyle groups, suggesting a generalized deterioration in psychological functioning in the later phases of the crisis. Similarly, Dooley et al. (2024) highlighted severe mental health concerns among Irish adolescents: over one-quarter described their mental health as "bad" or "very bad," with high rates of self-harm (39%), suicidal ideation (42%), and suicide attempts (11%) reported. Gender-diverse youth and females showed significantly higher vulnerability.

Khodami et al. (2022) conducted a global study to model psychological and quality-of-life trajectories under pandemic conditions. They found that perceived stress levels rose significantly and that difficulties in emotion regulation increased across the general population. Younger individuals and those who reported a decline in quality of life were particularly vulnerable. Importantly, their predictive model showed that extended quarantine duration was a key driver of increased stress and emotional dysregulation, suggesting that time under restrictions may be a primary risk factor for long-term psychological consequences.

In contrast, other studies painted a more stable or even positive picture. In Sweden, Kivi et al. (2021) examined older adults and found no significant declines in life satisfaction or other well-being indicators during the pandemic compared to pre-pandemic levels. Likewise, Peng and Roth (2022) found that among older U.S. adults, perceived health and satisfaction with their financial situation actually improved during lockdown, while loneliness remained largely unchanged.

Well-being was significantly influenced by daily behaviors and lifestyle factors during the pandemic. Zaninotto et al. (2022), in a 4-week longitudinal study in Italy and the UK, observed that well-being was initially shaped by internal factors such as affect and fatigue, but over time, external behaviors—like going outdoors, walking pets, or watching television—became more relevant. Importantly, frequent use of social media and video calls was associated with lower well-being. Similarly, Yang et al. (2020) found that emotional responses to online COVID-19 content varied depending on emotion regulation strategies, with cognitive reappraisal helping to reduce distress. Robbins et al. (2022) added that disrupted sleep, irregular routines, increased television use, and decreased physical activity were linked to higher levels of depression and anxiety among older adults, especially in women.

Other studies identified key sociodemographic predictors of well-being. Valiente et al. (2021), analyzing data from Spanish adults during lockdown, found that income level was the strongest predictor of well-being, followed by age, gender, prior mental health history, and COVID-related anxiety. Specifically, older adults demonstrated greater psychological resilience, while younger individuals and women reported lower well-being. Bhargav and Swords (2022) found that among Irish university students, financial concerns, chronic illness, and psychological distress significantly predicted elevated stress and reduced mental well-being.

At the intersection of emotion and cognition, social isolation was found to affect affective-cognitive integration. Bland et al. (2022), using tasks from the EMOTICOM neuropsychological test battery, showed that reduced social contact during isolation was associated with diminished positive bias in emotion recognition and attentional processing of emotional faces. Additionally, reduced frequency of contact with friends predicted lower cooperative behavior in economic decision-making tasks.

Finally, the role of hope, gratitude, and emotional meaning was not limited to emotional states, but also extended into well-being. Ang et al. (2022) found that hope and gratitude were robustly associated with lower psychological impact during COVID-19.

4 Conclusions

The SARS-CoV-2 pandemic has been a major global health crisis in recent times, profoundly affecting psychological and social functioning. For humans, few things can cause such an intense defensive emotional state as the threat of a virus. It is invisible, which prevents the detection of direct danger signals; uncertain, because its spread, evolution and consequences are beyond precise prediction; and uncontrollable, since its management depends largely on collective decisions, health measures, and factors beyond individual control. Moreover, in most cases, emergency measures caused a prolonged disruption of daily living conditions, activating multiple sources of stress—social isolation, economic uncertainty, work or educational overload, among others. The systematic review of 40 empirical studies examining the psychological effects of the COVID-19 pandemic reveals a complex and multifaceted impact on emotional states, cognitive functioning, and psychological well-being. While the pandemic was a globally shared stressor, the studies reviewed show substantial variability in individual responses, shaped by developmental stage, gender, coping strategies, cultural context, and temporal phase of the pandemic.

The studies reviewed reveal a complex landscape of affective reactivity during the pandemic. Negative emotional responses were more frequent and intense during the early phases, especially among youth, students, and those with maladaptive coping strategies. As the pandemic progressed, emotions of positive valence appeared. These findings suggest a degree of affective recovery as the pandemic progressed, possibly due to emotional habituation, increased coping, or adaptation to the "new normal." Positive emotional traits—such as gratitude, hope, and narrative coherence—acted as buffers, and over time, signs of emotional adaptation and post-traumatic growth became increasingly evident. The interaction between individual dispositions, coping resources, and socioenvironmental factors emerged as central to understanding emotional vulnerability and resilience during a global health crisis.

The reviewed studies collectively illustrate that the cognitive consequences of the pandemic were wide-ranging, extending from distorted memory and impaired appraisal to altered spontaneous thought and social cognition. Maladaptive appraisals—such as catastrophic thinking, fear-driven closure needs, and perceived helplessness—amplified distress and interfered with emotion regulation. In contrast, functional appraisals (e.g., challenge orientation, self-efficacy) reduced distress and promoted engagement. Studies on memory and spontaneous cognition (e.g., dreams, mind-wandering) revealed how pervasive the pandemic's emotional load was in shaping internal mental content, sometimes distorting affective memories or biasing attention. Emotion regulation strategies like cognitive reappraisal not only helped reduce immediate distress but were also associated with better long-term psychological health.

In terms of well-being, the evidence was more heterogeneous. While a subset of studies identified sharp declines in subjective well-being, especially among younger individuals, gender-diverse youth, and those with prior psychological vulnerabilities, other populations—most notably older adults—showed remarkable emotional stability. Protective factors such as routine, experimental diversity, meaningful activity, and social connection consistently emerged as buffers. Conversely, behavioral dysregulation—manifested in poor sleep, sedentary lifestyle, and high passive media consumption—was consistently linked to declines in emotional well-being.

However, several methodological and contextual limitations must be considered when interpreting the findings synthesized in this review. First, the studies included captured different phases of the pandemic, across distinct geographical and sociopolitical contexts, using varied measurement tools, sample characteristics, and data collection procedures. Public health measures—such as lockdowns, curfews, and restrictions on movement—varied widely across regions and timepoints, making it difficult to establish homogeneous conditions for comparison. As a result, the generalizability of the findings must be approached with caution.

Second, the very nature of the COVID-19 pandemic makes the use of control groups practically unfeasible. Given the absence of truly unaffected populations, most studies necessarily relied on descriptive and correlational methodologies. This limits the causal interpretability of the data and places greater emphasis on the need for cautious, context-sensitive interpretation of statistical associations.

Lastly, the majority of studies relied on self-report measures, which are inherently limited by subjectivity, social desirability, and introspective access. Few studies incorporated physiological or neurobiological indicators that could offer more objective markers of sustained stress or dysregulation (e.g., hair cortisol concentration, electrodermal activity). This reliance on subjective reports may obscure the full complexity of emotional and cognitive functioning under crisis conditions.

References

Alderfer, C. P. (1969). An empirical test of a new theory of human needs. *Organizational. Behavior and Human Performance, 4*(2), 142–175. https://doi.org/10.1016/0030-5073(69)90004-X

Almeida, D. M., Wethington, E., & Kessler, R. C. (2002). The daily inventory of stressful events: An interview-based approach for measuring daily stressors. *Assessment, 9*(1), 41–55. https://doi.org/10.1177/1073191102091006

Amodio, D. M. (2025). A learning and memory account of impression formation and updating. *Nature Reviews Psychology*. https://doi.org/10.1038/s44159-025-00445-x

Ang, J. Y.-Z., Monte, V., & Tsai, W. (2022). First-year college students' adjustment during the COVID-19 pandemic: The protective roles of hope and gratitude. *Translational Issues in Psychological Science, 8*(3), 375–388. https://doi.org/10.1037/tps0000320

Aristotle (2009). *Nicomachean ethics* (W.D. Ross & Brown, Trans.; Rev.ed.). Oxford University Press.

Aung, M. N., Komasawa, M., Nsereko, C., & Saito, K. (2025). Determinants maintaining healthcare personnel's motivation during COVID-19 in Uganda. *Scientific Reports*. https://www.nature.com/articles/s41598-025-86685-6

Barber, S. J., & Kim, H. (2021). COVID-19 worries and behavior changes in older and younger men and women. *The Journals of Gerontology: Series B, 76*(2), e17–e23. https://doi.org/10.1093/geronb/gbaa068

Barkham, M., Bewick, B., Mullin, T., Gilbody, S., Connell, J., Cahill, J., Mellor-Clark, J., Richards, D., Unsworth, G., & Evans, C. (2013). The CORE-10: A short measure of psychological distress for routine use in the psychological therapies. *Counselling and Psychotherapy Research, 13*(1), 3–13. https://doi.org/10.1080/14733145.2012.729069

Barrett, L. F., & Lida, T. (2024). Constructionist theories of emotions in psychology and neuroscience. In *Emotion Theory: The Routledge Handbook*. https://www.taylorfrancis.com/chapters/edit/10.4324/9781315559940-20

Barrett, L. F. (2025, January 23). *The theory of constructed emotion: More than a feeling*. https://doi.org/10.31234/osf.io/vwxup (preprint)

Bauman, Z. (2000). *Liquid modernity*. Polity Press.

Berwick, D. M., Murphy, J. M., Goldman, P. A., Ware, J. E., Jr., Barsky, A. J., & Weinstein, M. C. (1991). Performance of a five-item mental health screening test. *Medical Care, 29*(2), 169–176. https://doi.org/10.1097/00005650-199102000-00008

Bhargav, M., & Swords, L. (2022). Risk factors for COVID-19-related stress among college-going students. *Irish Journal of Psychological Medicine, 41*(2), 220–226. https://doi.org/10.1017/ipm.2022.33

Bland, A. R., Roiser, J. P., Mehta, M. A., Sahakian, B. J., Robbins, T. W., & Elliott, R. (2022). The impact of COVID-19 social isolation on aspects of emotional and social cognition. *Cognition and Emotion, 36*(1), 49–58. https://doi.org/10.1080/02699931.2021.1892593

Butler, J., & Kern, M. L. (2016). The PERMA-Profiler: A brief multidimensional measure of flourishing. *International Journal of Wellbeing, 6*(3), 1–48. https://doi.org/10.5502/ijw.v6i3.526

Browman, A. S. (2025). Centering social fit, self-concept fit, goal fit, and resource fit as core elements of students' experiences of belonging at school. *Educational Psychology Review*. https://link.springer.com/article/10.1007/s10648-025-10009-3

Caglayan-Akay, E., Ertok-Onurlu, M., & Komuryakan, F. (2025). Do the COVID-19 policies affect the happiness of the Turkish people? Evidence from the Life Satisfaction Survey. *International Journal of Community Well-Being*, 1–46. https://doi.org/10.1007/s42413-025-00234-7

Capurso, C. (2025). Editorial: Reviews in psychiatry 2023: Aging psychiatry. *Frontiers in Psychiatry, 16*, 1601909. https://doi.org/10.3389/fpsyt.2025.1601909

Carleton, R. N., Norton, M. A. P. J., & Asmundson, G. J. G. (2007). Fearing the unknown: A short version of the Intolerance of Uncertainty Scale. *Journal of Anxiety Disorders, 21*(1), 105–117. https://doi.org/10.1016/j.janxdis.2006.03.014

Castillo, J., Fan, H., Karaman, O. T., Shu, J., Stussi, Y., Kredlow, M. A., et al. (2024). Overestimating the intensity of negative feelings in autobiographical memory: Evidence from the 9/11 attack and Covid-19 pandemic. *Cognition and Emotion, 38*(7), 1048–1063. https://doi.org/10.1080/02699931.2024.2346757

Cohen, S., Kamarck, T., & Mermelstein, R. (1983). A global measure of perceived stress. *Journal of Health and Social Behavior, 24*(4), 385–396. https://doi.org/10.2307/2136404

Cregg, D. R., & Cheavens, J. S. (2021). Gratitude interventions: Effective self-help for mental health. *Clinical Psychology Review, 87*, 102033. https://doi.org/10.1016/j.cpr.2021.102033

Dattilo, T. M., Fisher, R. S., Traino, K. A., Roberts, C. M., Lehmann, V., Chaney, J. M., & Mullins, L. L. (2022). Dating anxiety during the global COVID-19 pandemic: Implications for college students. *Translational Issues in Psychological Science, 8*(3), 352–361. https://doi.org/10.1037/tps0000307

Deci, E. L., & Ryan, R. M. (2000). The "what" and "why" of goal pursuits: Human needs and the self-determination of behavior. *Psychological Inquiry, 11*(4), 227–268. https://doi.org/10.1207/S15327965PLI1104_01

Demiray, B., & Bluck, S. (2020). The relevance of remembered life events for the self in younger and older adults. *Self and Identity, 19*(4), 410–427. https://doi.org/10.1080/15298868.2019.1673752

Derogatis, L. R. (2000). *SCL-90-R: Administration, scoring, and procedures manual* (3rd ed.). Minneapolis, MN: National Computer Systems.

Diener, E., Emmons, R. A., Larsen, R. J., & Griffin, S. (1985). The Satisfaction With Life Scale. *Journal of Personality Assessment, 49*(1), 71–75. https://doi.org/10.1207/s15327752jpa4901_13

Diotaiuti, P., Valente, G., Mancone, S., Corrado, S., Bellizzi, F., Falese, L., et al. (2023). Effects of cognitive appraisals on perceived self-efficacy and distress during the COVID-19 lockdown: An empirical analysis based on structural equation modeling. *International Journal of Environmental Research and Public Health, 20*(7), 5294. https://doi.org/10.3390/ijerph20075294

Dooley, N., Power, E., Healy, H., Cotter, D., & Cannon, M. (2024). Mental health of Irish adolescents following the COVID-19 pandemic: Results from a population-based cross-sectional survey. *Irish Journal of Psychological Medicine, 41*(4), 430–438. https://doi.org/10.1017/ipm.2024.16

Dumciene, A., & Pozeriene, J. (2022). The emotions, coping, and psychological well-being in time of COVID-19: Case of master's students. *International Journal of Environmental Research and Public Health, 19*(10), 6014. https://doi.org/10.3390/ijerph19106014

Enav, Y., Iron, Y., Kivenson Bar-On, I., & Scharf, M. (2024). Protective factors against parental burnout during the COVID-19 pandemic. *Scientific Reports*. https://www.nature.com/articles/s41598-024-79020-y

Erikson, E. H. (1960). The problem of ego identity. In M. R. Stein, A. J. Vidich, & D. M. White (Eds.), *Identity and anxiety: Survival of the person in mass society* (pp. 37–87). Glencoe, IL: Free Press of Glencoe.

Felitti, V. J., Anda, R. F., Nordenberg, D., Williamson, D. F., Spitz, A. M., Edwards, V., Koss, M. P., & Marks, J. S. (1998). Relationship of childhood abuse and household dysfunction to many of the leading causes of death in adults: The Adverse Childhood Experiences (ACE) Study. *American Journal of Preventive Medicine, 14*(4), 245–258. https://doi.org/10.1016/s0749-3797(98)00017-8

Fernández-Abascal, E. G., & Díaz, M. D. M. (2022). Effects of the COVID-19 lockdown on personal well-being: A longitudinal study in Spanish population. *Anales de Psicología/Annals of Psychology, 38*(3), 458–468. https://doi.org/10.6018/analesps.509521

Fingerman, K. L., Ng, Y. T., Zhang, S., Britt, K., Colera, G., Birditt, K. S., & Charles, S. T. (2021). Living alone during COVID-19: Social contact and emotional well-being among older adults. *The Journals of Gerontology: Series B, 76*(3), e116–e121. https://doi.org/10.1093/geronb/gbaa200

Folkman, S., Lazarus, R. S., Dunkel-Schetter, C., DeLongis, A., & Gruen, R. J. (1986). Dynamics of a stressful encounter: Cognitive appraisal, coping, and encounter outcomes. *Journal of Personality and Social Psychology, 50*(5), 992–1003. https://doi.org/10.1037/0022-3514.50.5.992

Freud, S. (1915). The unconscious. In J. Strachey (Ed. & Trans.), *The Standard Edition of the Complete Psychological Works of Sigmund Freud* (Vol. 14, pp. 159–215). Hogarth Press.

Folville, A., Willems, S., Cheriet, N., Geurten, M., Guillemin, C., Muto, V., et al. (2023). Well-being during COVID-19-related first lockdown: Relationship with autobiographical memory and experiential diversity. *Applied Cognitive Psychology, 37*(5), 1059–1070. https://doi.org/10.1002/acp.4104

Giddens, A. (1991). *Modernity and self-identity: Self and society in the late modern age*. Stanford University Press.

Glickman, A. R., & La Greca, A. M. (2004). The Dating Anxiety Scale for Adolescents: Scale development and associations with adolescent functioning. *Journal of Clinical Child & Adolescent Psychology, 33*(3), 566–578. https://doi.org/10.1207/s15374424jccp3303_14

Gross, J. J. (2015). Emotion regulation: Current status and future prospects. *Psychological Inquiry, 26*(1), 1–26. https://doi.org/10.1080/1047840X.2014.940781

Habermas, T., & Köber, C. (2015). Autobiographical reasoning is constitutive for narrative identity: The role of the life story for personal continuity. In K. C. McLean & M. Pasupathi (Eds.), *The Oxford handbook of identity development* (pp. 149–165). Oxford University Press.

Hou, L., Long, F., Meng, Y., Cheng, X., Zhang, W., & Zhou, R. (2021). The relationship between quarantine length and negative affect during the COVID-19 epidemic among the general population in China: The roles of negative cognition and protective factors. *Frontiers in psychology, 12*, 575684. https://doi.org/10.3389/fpsyg.2021.575684

James, W. (1884). What is an emotion? *Mind, 9*(34), 188–205. https://doi.org/10.1093/mind/os-IX.34.188

Kahneman, D., Krueger, A. B., Schkade, D. A., Schwarz, N., & Stone, A. A. (2004). A survey method for characterizing daily life experience: The day reconstruction method. *Science, 306*(5702), 1776–1780. https://doi.org/10.1126/science.1103572

Kazak, A. E., Alderfer, M. A., Enlow, P. T., Lewis, A. M., Vega, G., Barakat, L. P., Kassam-Adams, N., Pai, A., Canter, K. S., & Hildenbrand, A. K. (2021). COVID-19 Exposure and Family Impact Scales: Factor structure and initial psychometrics. *Journal of Pediatric Psychology, 46*(5), 504–513. https://doi.org/10.1093/jpepsy/jsab026

Khodami, M. A., Hassan Seif, M., Sadat Koochakzadeh, R., Fathi, R., & Kaur, H. (2022). Perceived stress, emotion regulation and quality of life during the Covid-19 outbreak: A multi-cultural online survey. *Annales medico-Psychologiques, 180*(6), 514–518. https://doi.org/10.1016/j.amp.2021.02.005

Keltner, D., Tracy, J. L., Sauter, D., & Cowen, A. (2019). What basic emotion theory really says for the twenty-first century study of emotion. *Journal of Nonverbal Behavior, 43*(2), 195–210. https://doi.org/10.1007/s10919-019-00298-y

Kivi, M., Hansson, I., & Bjälkebring, P. (2021). Up and about: Older adults' well-being during the COVID-19 pandemic in a Swedish longitudinal study. *The Journals of Gerontology: Series B, 76*(2), e4–e9. https://doi.org/10.1093/geronb/gbaa084

Klaiber, P., Wen, J. H., DeLongis, A., & Sin, N. L. (2021). The ups and downs of daily life during COVID-19: Age differences in affect, stress, and positive events. *The Journals of Gerontology: Series B, 76*(2), e30–e37. https://doi.org/10.1093/geronb/gbaa096

Kumar, A. (2025). Unique models of embodied cognition and eco-social niches proposed to validate hypothesis of social attunement. *Frontiers in Psychiatry*. https://www.frontiersin.org/articles/10.3389/fpsyt.2025.1562061/full

Kyröläinen, A. J., & Kuperman, V. (2024). Emotional state of older adults during the COVID-19 pandemic: Insights from the Cognitive and Social Well-Being (CoSoWELL) corpus. *Experimental Aging Research, 50*(4), 482–505. https://doi.org/10.1080/0361073X.2023.2219188

Lawton, M. P., & Brody, E. M. (1969). Assessment of older people: Self-maintaining and instrumental activities of daily living. *The Gerontologist, 9*(3 Part 1), 179–186. https://doi.org/10.1093/geront/9.3_Part_1.179

LeDoux, J. E. (2012). Rethinking the emotional brain. *Neuron, 73*(4), 653–676. https://doi.org/10.1016/j.neuron.2012.02.004

LeDoux, J. E. (2021). What emotions might be like in other animals. *Current Biology, 31*(13), R821–R837. https://doi.org/10.1016/j.cub.2021.05.040

Low, R. S., Overall, N. C., Chang, V. T., Henderson, A. M., & Sibley, C. G. (2021). Emotion regulation and psychological and physical health during a nationwide COVID-19 lockdown. *Emotion, 21*(8), 1671. https://doi.org/10.1037/emo0001046

Lyotard, J.-F. (1979). *The postmodern condition: A report on knowledge*. Manchester University Press.

Marrazzo, J. (2025). Diverse metrics of success for physician-scientists: Reflections on my path and the evolving role of the National Institute of Allergy and Infectious Diseases. *The Journal of Infectious Diseases, 231*(2), 294–297. https://doi.org/10.1093/infdis/jiae495

Matić, T., Pregelj, P., Sadikov, A., & Rus Prelog, P. (2022). Depression, anxiety, stress, and suicidality levels in young adults increased two years into the COVID-19 pandemic. *International Journal of Environmental Research and Public Health, 20*(1), 339. https://doi.org/10.3390/ijerph20010339

Matos, M., McEwan, K., Kanovský, M., Halamová, J., Steindl, S. R., Ferreira, N., et al. (2021). The role of social connection on the experience of COVID-19 related post-traumatic growth and stress. *PLoS One, 16*(12), e0261384. https://doi.org/10.1371/journal.pone.0261384

Maslow, A. H. (1943). A theory of human motivation. *Psychological Review, 50*(4), 370–396. https://doi.org/10.1037/h0054346

McCaul, K. D., & Goetz, P. W. (n.d.). Worry. National Cancer Institute, Division of Cancer Control and Population Sciences. Retrieved September 28, 2025, from https://cancercontrol.cancer.gov/sites/default/files/2020-06/worry.pdf

Naaz, S. I., Hussein, R. M., Khan, H. B., Hussein, M. M., & Arain, S. A. (2022). Emotional responses and coping strategies of medical students during the COVID-19 pandemic. *Saudi Medical Journal, 43*(1), 61. https://doi.org/10.15537/smj.2022.43.1.20210698

Nitschke, J. P., Forbes, P. A., Ali, N., Cutler, J., Apps, M. A., Lockwood, P. L., & Lamm, C. (2021). Resilience during uncertainty? Greater social connectedness during COVID-19 lockdown is associated with reduced distress and fatigue. *British journal of health psychology, 26*(2), 553–569. https://doi.org/10.1111/bjhp.12485

Oerlemans, W. G. M., Bakker, A. B., & Veenhoven, R. (2011). Finding the key to happy aging: A day reconstruction study of happiness. *The Journals of Gerontology: Series B, 66*(6), 665–674. https://doi.org/10.1093/geronb/gbr040

Padmanabhanunni, A., & Pretorius, T. B. (2024). Fear of COVID-19 and PTSD: The protective function of problem-solving appraisals in mental health. *International Journal of Environmental Research and Public Health, 21*(2), 220. https://doi.org/10.3390/ijerph21020220

Page, M. J., McKenzie, J. E., Bossuyt, P. M., Boutron, I., Hoffmann, T. C., Mulrow, C. D., Shamseer, L., Tetzlaff, J. M., Akl, E. A., Brennan, S. E., Chou, R., Glanville, J., Grimshaw, J. M., Hróbjartsson, A., Lalu, M. M., Li, T., Loder, E. W., Mayo-Wilson, E., McDonald, S., et al. (2021). The PRISMA 2020 statement: An updated guideline for reporting systematic reviews. *BMJ, 372*, n71. https://doi.org/10.1136/bmj.n71

Peng, S., & Roth, A. R. (2022). Social isolation and loneliness before and during the COVID-19 pandemic: A longitudinal study of US adults older than 50. *The Journals of Gerontology: Series B, 77*(7), e185–e190. https://doi.org/10.1093/geronb/gbab068

Perasso, G., Serantoni, G., Paoletti, P., Maculan, A., & Lillo, C. (2023). Emotions, social support and positive resources during Covid19 pandemic: A qualitative-quantitative survey among italian adolescents. *Ricerche di psicologia: 1, 2023*, 83–105.

Perica, M. I., Ravindranath, O., Calabro, F. J., Foran, W., & Luna, B. (2021). Hippocampal-prefrontal connectivity prior to the COVID-19 pandemic predicts stress reactivity. *Biological Psychiatry Global Open Science, 1*(4), 283–290. https://doi.org/10.1016/j.bpsgos.2021.06.010

Pow, J., Lee-Baggley, D., & DeLongis, A. (2016). Threats to communion and agency mediate associations between stressor type and daily coping. *Anxiety, Stress, & Coping, 29*(6), 660–672. https://doi.org/10.1080/10615806.2015.1126258

Pow, J., Lee-Baggley, D., & DeLongis, A. (2017). Who is most likely to seek and give support in the face of agentic and communal threat? The roles of extraversion and agreeableness. *Journal of Research in Personality, 70*, 66–72. https://doi.org/10.1016/j.jrp.2017.04.006

Radloff, L. S. (1977). The CES-D Scale: A self-report depression scale for research in the general population. *Applied Psychological Measurement, 1*(3), 385–401. https://doi.org/10.1177/014662167700100306

Reitz, A. K. (2022). Self-esteem development and life events: A review and integrative process framework. *Social and Personality Psychology Compass, 16*(11), e12709.

Robbins, R., Weaver, M. D., Czeisler, M. É., Barger, L. K., Quan, S. F., & Czeisler, C. A. (2022). Associations between changes in daily behaviors and self-reported feelings of depression and anxiety about the COVID-19 pandemic among older adults. *The Journals of Gerontology: Series B, 77*(7), e150–e159. https://doi.org/10.1093/geronb/gbab110

Rogers, C. R. (1961). *On becoming a person: A therapist's view of psychotherapy*. Houghton Mifflin.

Rubin, D. C. (2022). *Autobiographical memory and the self in time*. Cambridge University Press. https://doi.org/10.1017/9781108635342

Russell, D. W. (1996). UCLA Loneliness Scale (Version 3): Reliability, validity, and factor structure. *Journal of Personality Assessment, 66*(1), 20–40. https://doi.org/10.1207/s15327752jpa6601_2

Scheier, M. F., Carver, C. S., & Bridges, M. W. (1994). Distinguishing optimism from neuroticism (LOT-R): Relations to anxiety, depression, and well-being. *Journal of Personality and Social Psychology, 67*(6), 1063–1078. https://doi.org/10.1037/0022-3514.67.6.1063

Seligman, M. E. P. (2011). *Flourish: A visionary new understanding of happiness and well-being*. Free Press.

Seppälä, E., Simon-Thomas, E., Brown, S. L., Worline, M. C., Cameron, C. D., & Doty, J. R. (2022). *The Oxford handbook of compassion science*. Oxford University Press.

Shchelkunov, S. N. (2009). How long ago did smallpox virus emerge? *Archives of Virology, 154*(12), 1865–1871. https://doi.org/10.1007/s00705-009-0536-0

Sherbourne, C. D., & Stewart, A. L. (1991). The MOS social support survey. *Social Science & Medicine, 32*(6), 705–714. https://doi.org/10.1016/0277-9536(91)90150-B

Sikka, P., Tuominen, J., Ezquerro Nassar, A., Kirberg, M., Loukola, V., Revonsuo, A., Valli, K., Windt, J., Bekinschtein, T. A., & Noreika, V. (2024). COVID-19 on mind: Daily worry about the coronavirus is linked to negative affect experienced during mind-wandering and dreaming. *Emotion, 24*(1), 177–195. https://doi.org/10.1037/emo0001255

Smith, B. W. et al. (2008). The Brief Resilience Scale: Assessing the Ability to Bounce Back from Stress. *International Journal of Behavioral Medicine, 15*(3), 194–200.

Taleb, N. N. (2007). *The Black Swan: The impact of the highly improbable*. Random House.

Titchener, E. B. (1909). *Lectures on the experimental psychology of the thought-processes*. The Macmillan Company. https://archive.org/details/lecturesonexperi00titciala

Tugade, M. M., Fredrickson, B. L., & Feldman Barrett, L. (2021). Positive emotions and emotional intelligence in the context of resilience. *Emotion Review, 13*(1), 15–25. https://doi.org/10.1177/1754073920951777

Valiente, C., Contreras, A., Peinado, V., Trucharte, A., Martínez, A. P., & Vázquez, C. (2021). Psychological adjustment in Spain during the COVID-19 pandemic: Positive and negative mental health outcomes in the general population. *The Spanish Journal of Psychology, 24*, e8. https://doi.org/10.1007/s10902-020-00352-3

van Tilburg, T. G., Steinmetz, S., Stolte, E., van der Roest, H., & de Vries, D. H. (2021). Loneliness and mental health during the COVID-19 pandemic: A study among Dutch older adults. *The Journals of Gerontology: Series B, 76*(7), e249–e255. https://doi.org/10.1093/geronb/gbaa111

Vanaken, L., Bijttebier, P., Fivush, R., & Hermans, D. (2022). Narrative coherence predicts emotional well-being during the COVID-19 pandemic: A two-year longitudinal study. *Cognition and Emotion, 36*(1), 70–81. https://doi.org/10.1080/02699931.2021.1902283

Watson, D., & Clark, L. A. (1999). The PANAS-X: Manual for the Positive and Negative Affect Schedule – Expanded Form. University of Iowa. https://www2.psychology.uiowa.edu/faculty/clark/panas-x.pdf

Weststrate, N. M., Jayawickreme, E., & Wrzus, C. (2022). Advancing a three-tier personality framework for posttraumatic growth. *European Journal of Personality, 36*(4), 704–725.

World Health Organization. (2020). *Coronavirus Disease (COVID-19) Pandemic*. WHO. https://www.who.int/emergencies/diseases/novel-coronavirus-2019

World Health Organization. (2024). *Impact of the COVID-19 pandemic on the rehabilitation workforce and service delivery in five countries of the WHO European Region*. WHO Regional Office for Europe. https://iris.who.int/bitstream/handle/10665/380535/WHO-EURO-2025-10489-50261-75807-eng.pdf

Wundt, W. (1897). *Outlines of psychology* (C. H. Judd, Trans.). Wilhelm Engelmann. https://archive.org/details/outlinesofpsycho00wunduoft

Xu, R. H., & Chen, C. (2024). Moderating effect of coping strategies on the association between the infodemic-driven overuse of health care services and cyberchondria and anxiety: Partial least squares structural equation modeling study. *Journal of Medical Internet Research, 26*(1), e53417. https://doi.org/10.2196/53417

Yan, D., Zhang, H., Guo, S., & Zeng, W. (2022). Influence of anxiety on university students' academic involution behavior during COVID-19 pandemic: Mediating effect of cognitive closure needs. *Frontiers in Psychology, 13*, 1005708. https://doi.org/10.3389/fpsyg.2022.1005708

Yang, Y., Liu, K., Li, S., & Shu, M. (2020). Social media activities, emotion regulation strategies, and their interactions on people's mental health in COVID-19 pandemic. *International Journal of Environmental Research and Public Health, 17*(23), 8931. https://doi.org/10.3390/ijerph17238931

Zhang, X., Mukherjee, A., Halassa, M. M., & Chen, Z. S. (2025). Mediodorsal thalamus regulates task uncertainty to enable cognitive flexibility. *Nature Communications, 16*(1), 2640. https://doi.org/10.1038/s41467-025-58011-1

Zaninotto, F., Bossi, F., Terry, P., Riccaboni, M., & Galli, G. (2022). The evolution of psychological and behavioral consequences of self-isolation during lockdown: A longitudinal study across United Kingdom and Italy. *Frontiers in Psychiatry, 13*, 826277. https://doi.org/10.3389/fpsyt.2022.826277

Zurlo, M. C., Cattaneo Della Volta, M. F., & Vallone, F. (2020). COVID-19 Student Stress Questionnaire: Development and Validation of a Questionnaire to Evaluate Students' Stressors Related to the Coronavirus Pandemic Lockdown. *Frontiers in Psychology, 11*, Article 576758. https://doi.org/10.3389/fpsyg.2020.576758

General Effects of the COVID-19 Pandemic on Cognition

Anja Podlesek and Voyko Kavcic

Abstract This chapter reviews the cognitive consequences of the COVID-19 pandemic, integrating findings on the direct effects of SARS-CoV-2 infection on cognition during the acute phase and in long COVID, and the broader cognitive repercussions of the pandemic in the general, non-infected population. The mechanisms underlying cognitive impairments in infected and non-infected individuals are presented, and the factors of individual differences in responses to the infection or chronic stress caused by the pandemic are discussed. The need for integrative, person-centred interventions that address both neurological and psychological factors contributing to cognitive consequences of the COVID-19 pandemic is emphasised.

Keywords COVID-19 pandemic · Post-acute sequelae of COVID-19 · Cognition · Stress · Individual differences

1 Introduction

The COVID-19 pandemic has left a profound and lasting impact on global health and psychological and cognitive functioning. As research on the pandemic consequences deepens, evidence is accumulating that the virus affects cognitive processes *directly*, via direct neuroinvasive effects of SARS-CoV-2 and associated neurological dysfunction, and *indirectly*, by altering the psychological environment of individuals and communities.

A. Podlesek (✉)
Department of Psychology, Faculty of Arts, University of Ljubljana, Ljubljana, Slovenia
e-mail: anja.podlesek@ff.uni-lj.si

V. Kavcic
Institute of Gerontology, Wayne State University, Detroit, MI, USA

M. Migueles Seco, A. Aizpurua Sanz (eds.), *Autobiographical Memory and the Covid-19 Pandemic*, https://doi.org/10.1007/978-3-032-03895-1_2

This chapter reviews the cognitive consequences of the COVID-19 pandemic, integrating findings on the direct effects of SARS-CoV-2 infection and the indirect cognitive impacts of pandemic-related stressors. We first examine the direct effects of COVID-19 on cognition, including deficits observed during the acute phase and in long COVID. Then we turn to the broader cognitive repercussions of the pandemic in the general, non-infected population, as accumulating evidence from cross-sectional and longitudinal studies suggests that pandemic-induced cognitive decline occurred even in uninfected groups, particularly among older adults, women, and individuals with pre-existing health or socioeconomic vulnerabilities. The mechanisms underlying cognitive impairments in infected and non-infected individuals are presented, and the factors of individual differences in responses to the infection or chronic stress caused by the pandemic are discussed, including age, illness severity, cognitive reserve, genetic vulnerabilities, social isolation, lifestyle changes, and emotional distress. Finally, we discuss the need for integrative, person-centred interventions addressing both neurological and psychological factors of the cognitive consequences of the COVID-19 pandemic.

2 Cognitive Changes During the COVID-19 Pandemic

2.1 The Direct Impact of COVID-19 on Cognition: Cognitive Changes in the Infected Population

Zhou et al. (2020) were among the first to report cognitive dysfunction in COVID-19 patients during the acute phase, particularly in sustained attention, which correlated with C-reactive protein levels. In a meta-analysis of 27 studies conducted up to the end of 2021, Crivelli et al. (2022) found impairments in visual perception, attention, executive functions, verbal fluency, naming, and memory among 2103 patients without prior cognitive deficits. Cohort studies showed that more than 50% of patients, especially those with more severe illness, reported moderate neurocognitive impairment, primarily in verbal fluency, attention, executive functions, and delayed recall. These deficits could be associated with sleep disturbances during the acute illness, as Jahrami et al. (2021) found an approximately 75% prevalence of sleep problems in COVID-19 patients, about twice the rate in general and healthcare populations. Those with severe COVID-19 symptoms, comorbidities, poorer psychological health, and hospitalisation experienced more severe cognitive impairments and faced a greater risk of lasting cognitive issues (Crivelli et al., 2022). In hospitalised patients without prior cognitive issues but with neurological symptoms during COVID-19, these symptoms were found to be risk factors for later deficits in processing speed, attention, memory, and executive function (Almeria et al., 2020).

While research on the cognitive effects during the acute phase of infection remains limited, the long-term consequences of the virus on cognitive functioning have been more extensively investigated. The longitudinal PROTECT study tracking 752 individuals aged 50+ before and during the pandemic concluded that domain-specific cognitive decline in COVID-19 patients resembled the decline in mild cognitive impairment, although with a slightly slower progression (Corbett et al., 2023). Faster than normal-ageing declines in executive function (small effect) and working memory (moderate effect) were found in COVID-19 patients during the first-year post-infection, the latter persisting into the second year. Taquet et al. (2022) concluded that while mood and anxiety disorders declined over time, elevated risk for cognitive impairments, dementia, psychosis, and epilepsy persisted 2 years post-infection (regardless of the virus variant) compared to other respiratory illnesses. Longitudinal studies are still needed to determine the trajectory and permanence of COVID-related cognitive impairments.

Individuals with a history of SARS-CoV-2 infection often experience long-lasting symptoms known as Long COVID or post-acute sequelae of COVID-19 (PASC). PASC typically begins within 3 months of disease onset and persists for at least 2 months without an alternative diagnosis (Soriano et al., 2022). However, symptom trajectories vary. They may persist, recur, or follow a remitting-relapsing pattern (Soriano et al., 2022). Most COVID-19 patients experience at least one lingering symptom (Carfi et al., 2020). About 80% report symptoms beyond 2 weeks (Diaz et al., 2021), and 5–30% report symptoms persisting over 9 months (Caspersen et al., 2022; Logue et al., 2021; Rajan et al., 2021). PASC can affect individuals of any age and is not necessarily tied to illness severity (Espinoza & Martella, 2025). Even individuals with mild disease report persistent symptoms. Symptoms usually worsen with physical or mental exertion (Davis et al., 2021). PASC significantly impairs quality of life and daily functioning (Carfi et al., 2020; Davis et al., 2021; Rajan et al., 2021; Soriano et al., 2022).

PASC symptoms are diverse. Alongside cough, shortness of breath, fatigue, post-exertional malaise, chest, joint and muscle pain, headache, mood disturbances (e.g., depression, anxiety), sleep disturbances, changes in smell and taste perception, nausea, and gastrointestinal issues, cognitive dysfunction is among the most common symptoms (Caspersen et al., 2022; Davis et al., 2021; Diaz et al., 2021; Espinoza & Martella, 2025; Groff et al., 2021; Nolen et al., 2022; Tabacof et al., 2022; Xie et al., 2022; Yelin et al., 2022; Zhou et al., 2020).

Studies have identified global and domain-specific cognitive impairments via self-report or neuropsychological testing. In PASC patients, self-reports indicated difficulties across nearly all cognitive domains (e.g., Caspersen et al., 2022; Davis et al., 2021; Nolen et al., 2022; Xie et al., 2022; Yelin et al., 2022). Davis et al. (2021) found that 88% of PASC individuals reported issues in attention, concentration, thinking, executive function (planning, organising, sequencing, abstracting), problem-solving, decision-making, and especially short-term memory. Al-Aly et al. (2021), analysing 73,435 medical records, observed increased neurocognitive complaints, fatigue, and 'brain fog[1]' within a month of infection. Up to one-third of

[1] The term 'brain fog' refers to subjective experiences of impaired focus, slowed processing, excessive effort needed for the thinking process, 'fuzziness', memory lapses, reduced mental clarity, dissociation, communication difficulties, and fatigue (McWhirter et al., 2023).

SARS-CoV-2 patients reported 'brain fog' 3 months or more post-infection (Fanshawe et al., 2025).

Despite compelling subjective complaints, PASC may go undetected by standard neuropsychological tools (Espinoza & Martella, 2025). Still, many studies support self-reported issues with objective findings. These studies vary in virus variants, vaccination rates, severity of COVID-19, sample size, demographics, clinical factors, presence of comorbidities, follow-up duration, and tools used to assess cognition and mental health, contributing to fragmented findings on the type, severity, and persistence of cognitive symptoms.

Overall, evidence from neuropsychological tests links PASC to global cognitive dysfunction (Ariza et al., 2022; Miskowiak et al., 2023), reduced processing speed (Ariza et al., 2022; Becker et al., 2021; Cavaco et al., 2023; Delgado-Alonso et al., 2022; Ferrucci et al., 2022; Poletti et al., 2022) and psychomotor deficits (Poletti et al., 2022). Memory impairment spans memory encoding and recall (Becker et al., 2021), long-term retention, episodic, working, prospective, and visuospatial memory (Caspersen et al., 2022; Cavaco et al., 2023; Delgado-Alonso et al., 2022; Ferruci et al., 2022; Llana et al., 2024). Language dysfunction, including verbal fluency, verbal learning, long-term verbal memory, and recall of appropriate words, has also been documented (Almeria et al., 2020; Ariza et al., 2022; Becker et al., 2021; De Tanti et al., 2023; Ferruci et al., 2022; Llana et al., 2024; Miskowiak et al., 2023). Further common impairments include attention and vigilance (Delgado-Alonso et al., 2022) and executive functions (Ariza et al., 2022; Becker et al., 2021; Delgado-Alonso et al., 2022; Miskowiak et al., 2023; De Tanti et al., 2023).

Meta-analyses (e.g., Ceban et al., 2022; Crivelli et al., 2022; Groff et al., 2021; Premraj et al., 2022) differ in effect sizes depending on their methods and focus. However, they converge in finding that a subset of individuals experiences mild to moderate impairments across multiple cognitive domains for a year or longer, especially in executive function (planning, inhibition, flexibility, abstract reasoning), verbal and visuospatial memory, attention (difficulty concentrating), processing speed, and language. Crivelli et al. (2022) found that patients without pre-existing cognitive issues had lower Montreal Cognitive Assessment (MoCA) scores (–0.94-point difference) than healthy controls up to 7 months post-infection, with impairment increasing with age. The most affected domains were executive function, attention, and memory. Fanshawe et al. (2025), analysing 66 high-quality studies, reported standardised mean differences between PASC patients and controls of −0.45 for executive function, −0.55 for learning and memory, −0.54 for complex attention, −0.54 for language, and −0.70 for perceptual motor function.

Cognitive deficits in PASC often persist for more than a year (Ferrucci et al., 2022), although gradual improvement over time has been observed (Cecchetti et al., 2022). Méndez et al. (2021) reported that about 60% of patients exhibited impairments in immediate verbal memory, learning, and fluency 2 months following hospital discharge. Similar rates of impairments in verbal learning and executive functions were found by Miskowiak et al. (2021) at 3–4 months, and Ferruci et al. (2022) at 5 months in slowed processing, with 25% showing deficits in delayed verbal recall, likely due to hypoxia. Schou et al. (2021) documented cognitive deficits in 25–78% of patients within 7 months of infection, including issues with attention, concentration, memory (short-term and general), encoding, language, verbal fluency, and praxis abilities, often co-occurring with fatigue, anxiety, depression, PTSD, and sleep problems. Liu et al. (2022) found that 12% of individuals aged 60 and above continued to experience cognitive impairment 1 year after infection.

2.2 *The Indirect Impact of the COVID-19 Pandemic on Cognition: Cognitive Changes in the Non-infected Population*

During the pandemic, changes in cognitive functioning were also reported among individuals who had not been infected. For some people without exposure to SARS-CoV-2, the pandemic itself acted as a trigger or exacerbating factor for cognitive and mental health difficulties. Governments introduced large-scale emergency measures, including self-isolation, economic shutdowns, restricted public services, and a shift to online education. Individuals faced home confinement, limited mobility, and behavioural adaptations such as physical distancing, mask-wearing, and remote work.

Numerous studies showed that pandemic-related stress and adaptation demands negatively affected mood (Xiao et al., 2020) and subjective well-being (Ammar et al., 2020b; Seckman, 2023) in the non-infected population. Sleep issues surged, affecting more than 30% of the general population and healthcare workers (Ammar et al., 2020b; Fu et al., 2020; Jahrami et al., 2021). Rates of anxiety, depression, distress, and post-traumatic stress symptoms rose (Fu et al., 2020; Liu et al., 2020; Xiong et al., 2020). Confinement intensified loneliness (Seckman, 2023), social isolation, sedentary behaviour, and poor dietary habits (Ammar et al., 2020a). These stressors—along with reductions in mobility, social interaction, physical activity, and cognitive stimulation—have long been recognised as contributors to cognitive decline. Research undertaken in the early months of the pandemic documented that stress likely had indirect effects on cognition in non-infected populations.

Many individuals uninfected with SARS-CoV-2 reported subjective cognitive complaints during the pandemic. Subjective cognitive complaints were associated with negative emotions, anxiety, depression, female gender, younger age, and underemployment (Fiorenzato et al., 2021; Podlesek et al., 2021). De Pue et al. (2021) also found that older adults self-reported poorer concentration and memory, closely tied to depression and loneliness. These findings suggest that cognitive complaints during the pandemic likely stemmed, at least in part, from emotional distress and stress-related responses.

Objective evidence of cognitive decline, however, varied across studies. Some researchers found no measurable impairments, or only minor ones, relative to the severity of the self-reported issues. Hausman et al. (2022), who tracked 189 older adults (aged 65–89) before and during 9 pandemic months, even observed slight improvements in objective cognitive performance and no changes in self-reported cognitive function. However, participants did report poorer sleep and physical health, reduced social engagement, and increased depression and apathy. Nogueira et al. (2021) compared the psychological functioning of 150 cognitively healthy Portuguese adults before and during the pandemic. Participants experienced increased depressive symptoms, reduced quality of life, and heightened feelings of loneliness and isolation. However, their cognitive performance remained largely stable. The rise in subjective cognitive complaints was closely linked to loneliness but did not correspond to actual cognitive decline on neuropsychological tests.

In contrast, several studies did find objective changes in cognitive function among uninfected individuals. Fiorenzato et al. (2021) reported objectively measured difficulties with attention, time orientation, and executive functions, while memory and verbal processing remained unaffected. Favieri et al. (2021) found that uninfected individuals with high post-traumatic stress symptoms related to COVID-19 showed executive deficits, performing worse on Stroop and Go/No-Go tasks. Schrempft et al. (2024) found that in older, mostly uninfected, non-dementia adults, subjective cognitive decline during the post-pandemic period was associated with both (1) poorer objective functioning (reduced ability to carry out daily tasks, increased fall risk, and lower well-being) and (2) additional stress-related factors, including lower income, health issues (multimorbidity, mobility issues, sensory impairment, appetite loss), health behaviours (former smoking, poor sleep), depressive symptoms, and increased social isolation since the onset of the pandemic. Notably, the association between subjective cognitive decline and objective cognitive performance was observed independently of depressive symptoms and demographic factors. These findings suggest that subjective cognitive decline cannot always be attributed solely to emotional distress.

Effects of Stressors Associated with the COVID-19 Pandemic on Cognition

Several studies have linked specific pandemic-related stressors to measurable changes in cognitive functioning during this period. One such stressor was uncertainty about the future. As an unprecedented global crisis, the COVID-19 pandemic was marked by unpredictable progression and limited personal control. Such uncontrollable situations provoke stress responses (Dickerson & Kemeny, 2004; Kemeny, 2003). In the early stages of the pandemic, uncertainty surrounding the virus's infectiousness and mortality amplified generalised anxiety and concerns related to health, financial stability, and personal safety (Droit-Volet et al., 2020). In our study, uncertainty about the future was a key factor behind university students' anxiety (Podlesek & Kavcic, 2021). Wu et al. (2022) found that during the pandemic, individuals experiencing high uncertainty were more likely to devalue future rewards and opt for immediate rather than delayed gains, reflecting a shorter time perspective. Their finding aligns with pre-pandemic research indicating that uncertainty can affect risk assessments and decision-making by shifting attention from goal-directed to habitual behaviours (Porcelli & Delgado, 2017) and leading to a present orientation and short-sighted decisions.

Another significant stressor was information overload. Constant exposure to pandemic-related news and misinformation intensified uncertainty, negative emotions, stress, and fear (Ying & Cheng, 2021) and contributed to higher rates of depression, anxiety (Fiorenzato et al., 2021; Xiong et al., 2020), and sleep disorders (Favieri et al., 2021). Overexposure through media and social networks also impacted cognitive functioning (Favieri et al., 2021). Xiao et al. (2022) reported that information overload, including mixed-quality content (the so-called infodemic), increased information anxiety, which led to impulsive buying in the early days of the pandemic. Excessive and unfiltered information disrupted cognitive reasoning, reduced attentional control, and made distinguishing true from fake information harder—ultimately promoting maladaptive coping.

A further significant stressor during the pandemic was confinement. Quarantine has been associated with a range of adverse outcomes, including exhaustion, insomnia, poor concentration, indecisiveness, and even post-traumatic stress, stemming from fear of infection, frustration, boredom, inadequate resources, and stigma (Brooks et al., 2020). Beyond infection-related quarantine, general confinement affected the broader population and posed significant risks to cognitive functioning. Lockdown measures (movement restrictions, school closures, online learning, and remote work) disrupted daily routines and sleep patterns, reducing focused activity. Fiorenzato et al. (2021) linked restricted mobility to declines in attention, time orientation, and executive functions, alongside increases in depression, anxiety, sleep issues, and changes in appetite and libido. Remote work and virtual collaboration further added to mental fatigue and cognitive load, lowering productivity (Kniffin et al., 2021). Blurred boundaries between work and leisure contributed to a distorted sense of time. People reported experiencing 'Blursday', a slowing of perceived time due to boredom and sadness (Chaumon et al., 2022; Droit-Volet et al., 2020).

COVID-19 restrictions increased social isolation, limiting the richness of social interactions and weakening social bonds. Reduced social contact emerged as a key predictor of cognitive deterioration. While Nogueira et al. (2021) found that perceived isolation in adults over 50 was linked only to subjective decline, other studies (e.g., Ingram et al., 2021) reported that objectively measured executive function worsened with reduced social contact, improving again as restrictions eased. Social isolation affects hormonal and immune function, disrupts brain connectivity, and impairs executive functions and memory, increasing long-term risk for neurological conditions like Alzheimer's disease (Bzdok & Dunbar, 2020). According to Holt-Lunstad et al. (2015), living alone, isolation, and loneliness also raise mortality risk by about 30%, even after adjusting for factors like age, sex, socioeconomic status, and health status.

Digital engagement helped counter social isolation during the pandemic. Internet use (e.g., social media, video calls) facilitated social connection. Li et al. (2022) found that, among adults aged 80 and older, use of information and communication technologies was linked to a lower risk of subjective cognitive decline. Similarly, Wieczorek et al. (2024) reported that digital literacy and participation in online social activities during the pandemic were associated with better immediate and delayed memory performance and self-rated memory. In a longitudinal study, Valla et al. (2025) found that older internet users without dementia maintained better psychological and cognitive health than non-users, suggesting that digital engagement may serve as a protective factor.

Pandemic-related disruptions negatively affected exercise and diet, both risk factors for dementia. In the PROTECT study, reduced physical activity predicted a faster executive function and working memory decline among non-infected individuals with mild cognitive impairment (Corbett et al., 2023). Home confinement led to less movement and increased sedentary time, harming both physical and mental health (Ammar et al., 2020a).

In sum, stress and mental health challenges during the pandemic contributed to elevated cognitive decline and dementia risk, especially among older adults (Schrempft et al., 2024), and also in other populations. Some longitudinal studies reported slight objective cognitive decline or increased mild cognitive impairment among uninfected individuals linked to pandemic conditions (Fernández Jiménez et al., 2023; Silva et al., 2022), although not always controlling for ageing. Other longitudinal studies demonstrated accelerated cognitive ageing. Amieva et al. (2022), who followed 15-year trajectories, found a significantly faster decline among French adults aged 80+ after the pandemic began. The UK PROTECT study (Corbett et al., 2023), tracking over 3100 individuals aged 50+, found a 50% faster decline in working memory and executive function, persisting into the second year of the pandemic. Those with mild cognitive impairment experienced an even steeper 60% acceleration. In dementia patients, about half showed worsened cognition and behaviour during the pandemic, with caregivers reporting decline in memory, attention, orientation, mood (depression, anxiety, agitation, irritability), and daily functioning (Borelli et al., 2021). Cognitively impaired individuals appeared especially vulnerable to stress, isolation, and lifestyle and healthcare disruptions. In contrast, Wang et al. (2025) found no evidence of pandemic-related objective cognitive decline among middle-aged nurses after controlling for age, education, and comorbidities, likely due to their higher socioeconomic status and health literacy, which may protect against neurocognitive decline (Vance et al., 2025).

3 Mechanisms of Cognitive Dysfunction Related to the COVID-19 Pandemic

3.1 Mechanisms of Cognitive Dysfunction During and After COVID-19 Infection

In the infected individuals, cognitive deficits such as deficits in behavioural inhibition and memory have been linked to acute COVID-19 symptoms like headache, sleep disturbance or insomnia (Cavaco et al., 2023; Delgado-Alonso et al., 2022), and anosmia/ageusia (Almeria et al., 2020; Cecchetti et al., 2022; Delgado-Alonso et al., 2022; Ferrucci et al., 2022; Llana et al., 2023), suggesting neurological origins of cognitive deficits after COVID-19.

Neurological symptoms may arise from viral neuroinvasion and immune dysregulation, including cytokine storms and neuroinflammation (Li et al., 2023). Inflammation can disrupt the blood–brain barrier, allowing cytokines into the central nervous system, activating microglia and oxidative stress, which promote cognitive decline.

Cognitive deficits were also linked to systemic and vascular inflammation during the acute phase of the infection (Ceban et al., 2022; Poletti et al., 2022) and brain vascular damage (Varga et al., 2020). Miskowiak et al. (2021) found that high D-dimer levels, indicating abnormal blood clot formation somewhere in the body, predicted poorer delayed verbal recall and psychomotor speed.

Other proposed mechanisms of cognitive dysfunction among infected individuals include elevated pro-inflammatory mediators (Mekhora et al., 2024), cortisol dysregulation, glutamatergic neurotoxicity, and structural brain changes, for example, grey/white matter alterations and reduced diffusivity (Díez-Cirarda et al., 2023; Paolini et al., 2023; Poletti et al., 2022). Brain connectivity changes have been observed with a multimodal imaging approach, including resting-state disruptions in the right frontal pole and the middle temporal gyrus (Paolini et al., 2023), and hypoconnectivity between parahippocampal, orbitofrontal, and cerebellar regions (Díez-Cirarda et al., 2023). Jiang et al. (2024) observed resting-state EEG abnormalities in COVID-19 patients, such as slowing of EEG, reduced alpha wave power, and epileptiform activity, resembling those observed in early stages of neurodegenerative diseases, particularly Alzheimer's disease and related dementia. Other mechanisms of PASC include coagulopathy, anti-neuronal antibodies, respiratory distress and hypoxia, especially affecting the hippocampus (Ferrucci et al., 2022; Foreman et al., 2025), demyelination, and thinner ganglion cell layers (Cavaco et al., 2023). Crivelli et al. (2022) also cite mechanisms such as cellular damage from viral invasion, secondary inflammatory responses, reduced ACE2 activity, sepsis, multi-organ damage, and delirium. Li et al. (2023) emphasise that multiple interacting mechanisms likely underlie COVID-19-related cognitive impairments.

Although the effects of SARS-CoV-2 on the central and peripheral nervous systems—and the resulting impact on cognitive functioning—are becoming increasingly well understood, the sheer number and variety of proposed neurophysiological mechanisms highlight that, so far, no unified theory explains how the SARS-CoV-2 virus impairs brain function. Additionally, a range of confounding factors—such as social isolation, intensive care unit stays, psychological distress, pre-existing conditions, fatigue, sleep disturbances, and baseline cognitive status—may further contribute to the cognitive dysfunction observed in PASC (Cavaco et al., 2023; Whiteside et al., 2022).

Many studies suggest that fatigue and mental health issues might partly contribute to cognitive complaints in PASC patients. Subjective reports of 'brain fog' frequently co-occur with symptoms of depression, anxiety, and distress (Van der Feltz-Cornelis et al., 2024). Meta-analyses indicate higher rates of depression, anxiety, post-traumatic stress symptoms and sleep disorders among infected individuals compared to general or healthcare populations, even beyond 6 months post-infection (Deng et al., 2021; Groff et al., 2021; Krishnamoorthy et al., 2020; Poletti et al., 2022; Premraj et al., 2022). Ceban et al. (2022) found that one-third of PASC patients suffered from persistent fatigue (lack of energy, muscle weakness, malaise), accompanied by marked functional impairment lasting 12 or more weeks, with similar prevalence in both hospitalised and non-hospitalised groups. In a meta-analysis, Foreman et al. (2025) controlled for potential confounders and found that higher levels of anxiety and depression predicted slightly lower cognitive performance. Knapp et al. (2024) found no clinically significant objective cognitive impairments in COVID-19 patients without severe medical or psychiatric comorbidities. Austin et al. (2024) proposed that objective cognitive decline may result from a combination of neuroinflammation, pain, and mood disturbances. Symptoms like depression, anxiety, stress, insomnia, and fatigue may reduce cognitive test performance, complicating efforts to distinguish the direct neurological effects of COVID-19 from indirect psychological and physiological factors. The relationship between mood and objective cognitive performance in PASC patients remains unclear (Austin et al., 2024).

Notably, subjective cognitive complaints following COVID-19 do not always align with objective cognitive deficits. Several studies have reported inconsistent correlations between self-reported everyday memory or cognitive concerns and performance on neuropsychological tests (e.g., Almeria et al., 2020; Costas-Carrera et al., 2024; Miskowiak et al., 2021). For example, Cavaco et al. (2023) found that while 33% of COVID-19 patients reported cognitive concerns, only 20% demonstrated measurable impairments on testing. This pattern mirrors findings in other post-viral syndromes. Johnston et al. (2013), for instance, reported a 3.28% prevalence of self-reported post-viral fatigue, compared to only 0.76% based on clinical diagnosis. Correlations between subjective complaints and objective test scores in PASC patients are typically low (Ariza et al., 2022; Cavaco et al., 2023; Whiteside et al., 2022), although some studies (Miskowiak et al., 2021) have found moderate to high associations. Further complicating this picture, studies such as Almeria et al. (2020) observed no significant differences in test performance between patients with and without complaints but noted that those reporting complaints scored higher on measures of anxiety and depression.

While subjective memory complaints without corresponding objective impairment have been identified as an early marker of accelerated cognitive decline preceding mild cognitive impairment and dementia (Warren et al., 2022), pre-pandemic research has shown that such complaints, particularly in older adults, can also reflect psychological factors such as depression, anxiety, chronic stress, low sense of control, reduced self-esteem, and high neuroticism (Comijs et al., 2002; dos Santos et al., 2012; Steinberg et al., 2013). Additionally, subjective cognitive complaints in working adults were frequently linked to sleep disturbances, stressful life events, emotional exhaustion, burnout, physical illness, and diminished executive functioning (Jacob et al., 2019; Stenfors et al., 2013). Moreover, the phenomenon of diagnosis threat—where awareness of PASC increases symptom reporting—may further exacerbate cognitive complaints (Winter & Braw, 2022). These findings suggest that discrepancies between subjective and objective cognitive decline in PASC may, at least in part, be attributable to mental health factors and stress vulnerability. In some cases, such discrepancies have led clinicians to classify PASC-related cognitive complaints as functional cognitive disorders (Becker et al., 2023). This perspective often results in treatment approaches limited to psychiatric or psychological care, rather than comprehensive neurocognitive rehabilitation. However, growing evidence supports the view that post-COVID cognitive impairments often have a neurological basis, may persist in the long term for some individuals, and are not solely psychiatric or psychosomatic in origin (Austin et al., 2024; Espinoza & Martella, 2025).

Several studies have identified psychosocial stress as a significant risk factor for the development and persistence of PASC. For example, Gebhard et al. (2024) linked the higher incidence of PASC among female COVID-19 outpatients to the disproportionate psychosocial burdens women faced during the pandemic, including parenting, home-schooling, caregiving, and financial and personal security concerns. Wang et al. (2022) found that elevated levels of depression, anxiety, COVID-related worry, perceived stress, and loneliness—measured early in the pandemic, before SARS-CoV-2 infection—predicted increased risk for developing post-COVID-19 conditions and experiencing impairments in daily functioning. Similarly, Milde et al. (2023) reported that psychosomatic symptom burden, fear of COVID-related health consequences, perceived chronic stress, and depression predicted higher odds and severity of post-COVID symptoms. These findings suggest that the development and persistence of PASC symptoms may be influenced by pre-existing psychosocial conditions and chronic stress, conceptualised through the framework of allostatic load—the cumulative physiological burden of repeated or chronic stress exposure. Allostatic load manifests in dysregulation across multiple physiological systems, including the hypothalamic-pituitary-adrenal (HPA) axis, immune and inflammatory pathways, and cardiovascular and metabolic functions. It is associated with measurable changes in biomarkers (e.g., blood pressure, C-reactive protein) and may contribute to adverse health outcomes (Gómez-García et al., 2025). It may also contribute to the amplification or prolongation of the existing neurological dysfunction and hinder cognitive recovery. Psychological distress and mental health conditions such as anxiety and depression may thus not only arise as consequences of PASC but also potentiate neurobiological dysregulation, forming a feedback loop that exacerbates cognitive and emotional symptoms. Nevertheless, evidence is not entirely consistent. Fazekas et al. (2025) found that ongoing COVID-19 symptoms were predicted by pre-existing perceived stress but not by objective markers of chronic stress (e.g., number of stressful life events, hair cortisol). This raises the possibility that some associations reflect a shared tendency to report both stress and PASC symptoms, underscoring the need for longitudinal designs with validated biological stress markers.

Taken together, the current body of evidence suggests that cognitive dysfunction in PASC is underpinned by multifactorial mechanisms involving complex interactions between neurological, psychological, and physiological factors. The absence of a singular explanatory model underscores the need for an integrative biopsychosocial approach that accounts for both the direct biological effects of SARS-CoV-2 and indirect influences such as mental health, pre-existing biological and psychological vulnerabilities, and social context. Becker et al. (2023) advocate for such a comprehensive framework, which includes evaluation of premorbid functioning and environmental stressors. This multidimensional perspective is crucial not only for accurate diagnosis but also for developing individualised, holistic interventions tailored to the diverse needs of PASC patients.

3.2 Mechanisms of Cognitive Deficits Arising from Pandemic-Related Circumstances

Changed cognitive functioning among individuals not infected with SARS-CoV-2 was likely a consequence of the pandemic-related stress. Stress—defined as a state of emotional strain resulting from a real or perceived threat to homeostasis, when environmental demands exceed a person's ability to cope (James et al., 2023; Selye, 1956)—has long been recognised as a significant factor influencing mood and emotions, cognition, behaviour, well-being, and health. It exerts multifaceted effects on physiological systems, including the immune, cardiovascular (e.g., elevated heart rate, oxygen demand, blood pressure), gastrointestinal (e.g., altered appetite and digestion), and endocrine systems (Yaribeygi et al., 2017). Emotionally, stress often triggers negative affect such as tension, irritability, sadness, and guilt, although it may occasionally evoke positive affect, such as enthusiasm (Zhaoyang et al., 2020). Cognitively, stress can lead to mental slowness, confusion, forgetfulness, poor concentration, intrusive thoughts, and persistent worry. Severe stress may further impair learning, memory, decision-making, and abstract reasoning. It promotes a shift from flexible, goal-directed behaviours to more rigid, habitual, and cognitively less demanding responses (Kemeny, 2003; Porcelli & Delgado, 2017; Sandi, 2013; Schwabe & Wolf, 2013). Excessive arousal may also undermine performance by making routine actions conscious and slower or inducing panic and freezing cognition (Bourne Jr & Yaroush, 2003). Moreover, stress has been associated with decreased social behaviour and increased aggression (Sandi & Haller, 2015).

Pre-pandemic studies found that stress significantly modulates cognitive processes, including attention, memory, executive functioning, and social cognition (Diamond et al., 2007; Sandi & Haller, 2015). Acute stress activates the sympathetic adrenal medullary system and the HPA axis, increasing the release of glucocorticoids like cortisol into the bloodstream (James et al., 2023; Kemeny, 2003). While short-term elevations in cortisol can be adaptive—enhancing focus on a single, salient stimulus (usually the stressor) to help the individual respond effectively to immediate challenges (Liston et al., 2009; Shields et al., 2016) as well as enhancing memory encoding and consolidation (Roozendaal et al., 2009; Harlé et al., 2013; Henckens et al., 2009; McGaugh, 2015)—acute stress may also disrupt memory retrieval, working memory, and cognitive flexibility (Qin et al., 2009; Shields et al., 2016). High-intensity stress frequently impairs reasoning, learning, problem-solving, and retrospective and prospective memory (Bourne Jr & Yaroush, 2003; Roozendaal et al., 2009; Schwabe et al., 2012).

Chronic (persistent or recurring) stress, characterised by ongoing and unresolved challenges, exerts a more detrimental influence on neural function and structure (James et al., 2023). Prolonged cortisol elevation affects neuronal excitability and reshapes emotional and cognitive functioning, with structural and functional changes in brain regions like the hippocampus (linked to memory deficits); amygdala (linked to emotional reactivity), and prefrontal cortex, rich in glucocorticoid receptors (linked to working memory and fear regulation) (Chattarji et al., 2015; James et al., 2023). Chronic stress damages hand–eye coordination, attention shifting, explicit memory, executive function, and impulse control by altering brain connectivity, and is linked to neurodegenerative and psychiatric disorders (James et al., 2023; Liston et al., 2009; Sandi, 2013; Shields et al., 2016). Stress exposure during critical developmental periods increases the risk of long-term mental and physical health issues (James et al., 2023). Self-perceived stress and negative life experiences are associated with global cognitive performance (Leng et al., 2013). Encouragingly, some stress-induced impairments may reverse once the stressor is removed (Liston et al., 2009).

Chronic stress also compromises immune function and increases risks for cardiovascular disease, ulcers and psychiatric conditions like anxiety, post-traumatic stress disorders, and depression (Kemeny, 2003; Schneiderman et al., 2005; Thorsteinsson et al., 2019). It contributes to fatigue, marked by decreased vigilance, irritability, low energy, decreased motivation, disengagement and reduced performance, and to poorer self-rated health, chronic pain, fibromyalgia, and other functional somatic disorders (James et al., 2023; Kocalevent et al., 2011; Taylor & Dorn, 2006).

Stress and elevated glucocorticoid release are also strongly associated with disrupted sleep (Van Reeth et al., 2000). Poor sleep—whether shorter, shallower or fragmented—has been linked to emotional dysregulation, heightened anxiety symptoms, and impaired cognitive function (Altena et al., 2016; Baglioni et al., 2016; Cousins et al., 2018; Reddy et al., 2017). Many studies showed that inadequate sleep affects attention, working memory, and executive functions (Scullin & Bliwise, 2015). Sleep deprivation reduces vigilance and impairs performance in tasks involving memory encoding, learning, logical reasoning, and decision-making (Alhola & Polo-Kantola, 2007; Krueger, 1989). Even partial sleep loss can hinder declarative learning (Cousins et al., 2018), likely due to reduced alertness, attentional lapses, and slower responses (Alhola & Polo-Kantola, 2007).

Given the well-documented effects of chronic stress on cognition, health, and sleep, the widespread cognitive changes observed during the COVID-19 pandemic can plausibly be attributed to prolonged stress exposure.

Although individuals without confirmed SARS-CoV-2 infection may have experienced less pronounced cognitive impairment than those who were infected, the overall patterns of cognitive disruption often mirrored each other (see Table 1). Moreover, infected individuals were likely subject to both the physiological consequences of the virus and the psychological stressors associated with the pandemic.

Table 1 Comparison of direct and indirect cognitive effects of the COVID-19 pandemic

Dimension	Direct effects of SARS-CoV-2 infection	Indirect effects of pandemic-related stress
Population affected	Infected individuals, including those with mild, severe, or PASC symptoms	Non-infected individuals affected by stressors such as lockdowns, isolation, and lifestyle disruption
Onset and duration	Onset during the acute phase or within 3 months post-infection (PASC); symptoms may persist for more than 12 months	Gradual onset due to prolonged psychosocial stressors; variable persistence across individuals
Subjective cognitive complaints	'Brain fog', attention/concentration issues, slowed processing, memory lapses, word-finding difficulty	Similar complaints: forgetfulness, distractibility, reduced concentration, mental fatigue, 'Blursday'
Objective cognitive deficits	Confirmed impairments in attention, executive function, memory (working, episodic, prospective), processing speed, and language	Mixed findings: some studies show no measurable deficits, others report declines in attention, executive function, and working memory
Mechanisms	Neuroinflammation, cytokine storm, blood–brain barrier disruption, hypoxia, vascular damage, grey/white matter changes, altered EEG patterns, neurodegeneration, demyelination	HPA axis activation, elevated cortisol, altered sleep, emotional distress, social isolation, reduced stimulation, uncertainty, information overload and 'infodemic'
Neural substrates	Hippocampus, prefrontal cortex, amygdala, parahippocampal and orbitofrontal regions, middle temporal gyrus, cerebellum	Hippocampus, prefrontal cortex, amygdala, stress-induced functional connectivity
Psychiatric comorbidities	High rates of fatigue, depression, anxiety, PTSD, sleep disturbance; mood often interacts with cognitive complaints	Emotional distress (e.g., anxiety, depression, burnout), loneliness, lower well-being, and increased sleep disruption
Confounding factors	ICU stays, severity of illness, comorbidities, oxygen levels, age, pre-existing psychiatric or neurological conditions	Social isolation, income loss, digital fatigue, uncertainty, media overexposure, gender, age, caregiving burden
Modifying factors	Severity of illness, age, cognitive reserve, olfactory symptoms, insomnia, inflammation markers (e.g., D-dimer), APOEε4 gene, pre-existing health status, vaccination, variant type	Age, sex, socioeconomic status, neuroticism, psychological resilience, health literacy, lifestyle, social isolation, stress perception, digital engagement
Age-related vulnerability	Older adults are more likely to experience persistent deficits resembling dementia	Young adults report more mental health symptoms; older adults show greater emotional stability and resilience
Sex differences	Mixed evidence; higher prevalence of PASC in females	Women more frequently report cognitive complaints, greater emotional distress, and more adaptive coping

(continued)

Table 1 (continued)

Dimension	Direct effects of SARS-CoV-2 infection	Indirect effects of pandemic-related stress
Evidence of reversibility	Some recovery over time; deficits may persist >12 months, particularly in older or severely affected individuals	Stress-related cognitive impairments typically reverse as stressors are removed; individual outcomes depend on resilience and ongoing stress levels
Interventions needs	Thorough cognitive assessment; cognitive rehabilitation, memory/attention training, mood management, sleep/nutrition interventions, workplace accommodations, neurostimulation	Psychological and lifestyle interventions: promote sleep hygiene, stress regulation, physical and mental activity, social engagement, resilience-building; support access to digital tools and mental healthcare

As the virus spread, distinguishing between the cognitive effects of direct viral infection and those arising from pandemic-related psychological and environmental stressors became increasingly challenging.

4 Individual Differences in Cognitive Dysfunction Related to the COVID-19 Pandemic

The cognitive impact of the COVID-19 pandemic exhibited considerable inter--individual variability. While some individuals experienced significant cognitive decline, others appeared to remain relatively unaffected.

4.1 *Individual Differences in Post-COVID Cognitive Impairments*

In the infected individuals, some studies found persistent cognitive problems independent of sociodemographic characteristics, pre-existing health status, or mood symptoms (Becker et al., 2021; Davis et al., 2021; Hampshire et al., 2021). Others linked (subjective) cognitive decline to acute symptoms (e.g., headache, sleep disturbances), as well as female gender, age, lower education, anxiety, depression, fatigue and reduced quality of life (Cavaco et al., 2023; Mazza et al., 2021).

Severity of illness. Most studies reported higher prevalence of post-infection cognitive problems in hospitalised or severe cases (Becker et al., 2021; Davis et al., 2021; Liu et al., 2022; Vakani et al., 2024), as well as a positive association between cognitive impairments and the severity of initial illness (Austin et al., 2024; Foreman et al., 2025; Hampshire et al., 2021), and long-term hyposmia (Cristillo et al., 2021).

However, some studies (e.g., Logue et al., 2021) reported that cognitive impairments also occurred regardless of acute COVID-19 phase severity.

Age. Foreman et al.'s (2025) meta-analysis demonstrated a small-to-moderate negative association between age and cognitive performance in PASC patients. Gonzales Aleman et al. (2025) further found that PASC phenotypes were age-dependent: while young and middle-aged adults showed infrequent and mild attentional impairments, older adults exhibited dementia-like deficits in memory, language, and executive function. The risk of moderate-to-severe impairment doubled in older adults, particularly in cases with severe infection and olfactory symptoms. In children, Taquet et al. (2022) observed elevated risks of neurological complications (e.g., seizures, encephalitis), but not mood or anxiety disorders. Cognitive symptoms in children were generally transient, unlike in adults, where such symptoms tended to persist.

Genetic and hormonal influences. In a meta-analysis of 3069 infected individuals aged 75 and older, Acharya et al. (2023) identified associations between declines in memory and global cognition and the presence of APOEε4 and TMPRSS11D genes, both linked to SARS-CoV-2 activation and immune response. These cognitive declines were likely due to disruptions in the blood–brain barrier, neuroplasticity, and systemic immune responses. APOEε4 is recognised as one of the primary genetic risk factors for the development of Alzheimer's disease. Recent findings suggest that individuals with the ε4 genotype were more likely to be infected with COVID-19, whereas the ε2 allele showed a protective effect. APOE gene polymorphism may therefore play an important role in vulnerability for COVID-19 and cognitive decline (Tavares-Júnior et al., 2023).

Cognitive reserve. Individual differences in cognitive reserve—the adaptability of cognitive processes that helps to explain differential susceptibility of cognitive abilities or day-to-day function to brain ageing, pathology or insult (Stern et al., 2020)—have influenced the severity of cognitive deficits in PASC. De Tanti et al. (2023) found that, 8 months after discharge from a rehabilitation hospital, severely affected COVID-19 patients with higher cognitive reserve showed better executive functioning. The authors suggested that cognitive reserve may facilitate reintegration into normal social activities post-discharge. Similarly, Foreman et al. (2025) showed that cognitive reserve—measured through proxies such as educational level, validated cognitive reserve instruments, premorbid IQ, and occupational complexity—moderated the relationship between COVID-19 severity and post-COVID cognitive dysfunction. The strongest moderating effects were observed in delayed recall and visual processing speed. Higher cognitive reserve was associated with smaller cognitive deficits following COVID-19, whereas lower cognitive reserve was linked to greater impairments.

Batty et al. (2020), using UK Biobank data, found that individuals with lower cognitive performance, poorer psychological indicators (e.g., psychological distress, mental health, neuroticism), and socio-economic disadvantage 10–14 years pre-pandemic were more likely to be hospitalised for COVID-19. After controlling for confounders, only cognitive performance remained a significant predictor of hospitalisation, suggesting that prior cognitive reserve and health literacy might have contributed to protective health behaviours and influenced the course of illness.

4.2 *Individual Differences in Stress-Related Cognitive Impairments*

Substantial variability in the effects of the pandemic-related stressors was also observed among non-infected individuals. This heterogeneity is consistent with pre--pandemic research, which highlighted a range of individual differences in cognitive responses to stress. Such differences have been attributed to factors including the appraisal of stressors (i.e. whether they are perceived as threats or challenges), traits like self-confidence and self-esteem, resilience, perceived control, cognitive reserve, coping strategies, social skills, and the level of social support (Farrer et al., 2016; James et al., 2023; Uchino, 2009).

Biological and genetic moderators of the stress-cognition relationship have also been identified. These include the presence of genetic markers such as the S allele of the serotonin transporter gene (Nair et al., 2020a, 2020b) and the ApoE-ε4 allele (Montoliu et al., 2018), as well as hormonal influences such as cortisol levels, which vary with age and sex (James et al., 2023). Increased susceptibility to stress has been associated with younger age, lower socioeconomic status, female gender (Shields et al., 2016), and poorer self-rated health (Kocalevent et al., 2011; Novais et al., 2017; Valerio et al., 2016). These same risk factors were confirmed during the pandemic, with cognitive and mental health changes linked to demographic variables (e.g., age, gender), underemployment and remote work (Fiorenzato et al., 2021; Pieh et al., 2021; Podlesek et al., 2021; Santangelo et al., 2021), income (Pieh et al., 2021), health status, personality traits such as neuroticism (Podlesek et al., 2021), and psychological resilience (Zager Kocjan et al., 2021). Additional contributing factors included frequent exposure to COVID-related news, pre-existing chronic or psychiatric conditions, use of social media for pandemic-related content, and living in high-risk areas (Fiorenzato et al., 2021; Xiong et al., 2020). The following sections explore some of these factors in greater detail.

Sex. During the COVID-19 lockdowns, women more frequently reported subjective cognitive issues like attention deficits and memory lapses (Fenollar-Cortés et al., 2021; Fiorenzato et al., 2021; Guidice et al., 2022; Hsu et al., 2024; Maggi et al., 2022), although objective cognitive performance data showed few sex-based differences. Numerous studies observed that women also experienced more stress, anxiety, and depression than men, particularly during the early phases of the pandemic (Pieh et al., 2021). They were also more likely to express negative emotions such as anxiety and anger (Maggi et al., 2022). In a Spanish longitudinal study, women began lockdown with greater emotional distress (e.g., avoidance, intrusive thoughts) but showed more emotional recovery over time than men (Fenollar-Cortés et al., 2021). Similarly, an Italian study found that women were more concerned about COVID-19, which impacted their mood more intensely than men (Guidice et al., 2022). Women were also more likely to perceive the pandemic as a serious health threat, comply with health restrictions, and adopt hygiene practices (Attanasi et al., 2021; Galasso et al., 2020). In terms of coping, women favoured active strategies like seeking social support or practising mindfulness, whereas men tended toward emotion-focused approaches such as avoidance or substance use (Reisch et al., 2021). These differences may stem from broader societal roles, caregiving duties, emotional regulation styles, and risk perceptions.

Age. Several studies reported more severe mental health issues among younger adults during the pandemic (Maggi et al., 2022; Pieh et al., 2021; Podlesek & Kavcic, 2021). A meta-analysis by Li et al. (2021) found that rates of depression and anxiety among college students reached 39% and 36%, respectively. Younger individuals were also more vulnerable to the negative effects of restrictions, showing both increased subjective complaints (Fiorenzato et al., 2021) and declines in objective performance (Menze et al., 2022). In contrast, middle-aged adults exhibited lower levels of anxiety, suggesting greater emotional stability during lockdowns. Similarly, older adults reported less pandemic-related stress and demonstrated higher emotional stability, well-being, and coping capacity (Seckman, 2023).

Lifestyle and health. Prior physical and mental health conditions, lifestyle factors, and levels of intellectual and social engagement significantly influenced cognitive outcomes during the pandemic. Nogueira et al. (2021) found that individuals with higher pre-COVID-19 quality of life and a strong sense of well-being experienced fewer difficulties with emotion regulation, depression, anxiety, loneliness, and cognitive complaints during the pandemic. The PROTECT study (Corbett et al., 2023) linked declines in executive function and working memory during the first year of the pandemic to reduced physical activity, increased alcohol use, depression and loneliness. In older adults, maintaining daily movement, structured exercise or regular walking was associated with slower cognitive decline and lower depression levels (Harb et al., 2024), while decreased walking speed was identified as a significant predictor of new cognitive complaints (Tanaka et al., 2023). Regular physical activity and quality sleep were linked to better working memory performance and more positive self-perceptions of memory abilities in everyday life (Guardia et al., 2024). Overall, these findings underscore the critical role of lifestyle behaviours and mental health in maintaining cognitive function, with unhealthy habits and negative emotional states emerging as key risk factors for accelerated cognitive decline.

Resilience. Resilience, defined as the capacity to maintain or restore mental well--being in the face of adversity (Herrman et al., 2011), has emerged as particularly important during the COVID-19 pandemic. Studies have shown that lower levels of resilience were associated with more pronounced symptoms of depression, anxiety, and anger, factors closely linked to increased subjective cognitive complaints among older adults (Maggi et al., 2022; Miller et al., 2021; Santangelo et al., 2021). In line with this, Oyejide et al. (2025) reported that the group of participants who complied with the lockdown measures had lower anxiety and depression, and better memory recall performance compared to those who were in the sufferer or defiant groups.

Social engagement and support. In times of crisis or distress, social capital—reflected in the strength of one's social network and the frequency of received support—plays a crucial role in fostering resilience (Bzdok & Dunbar, 2020). Even prior to the pandemic, research consistently demonstrated that responses to stress were shaped by a range of social factors, including social support (Cohen, 2004), social skills (Uchino, 2009), and broader social capital (Xiao et al., 2020). Cohen (2004) identified three core components of social support: instrumental (material aid), informational (guidance or advice), and emotional (empathy and reassurance). With the onset of the COVID-19 pandemic, opportunities for communication and interpersonal connection diminished, leading to a widespread decline in social well--being. Reduced access to social support and heightened loneliness were particularly detrimental for older adults, emerging as key risk factors for cognitive decline (Harb et al., 2024; Hwang, 2024; Tanaka et al., 2023). However, those who used technology to maintain contact with family and friends were less likely to experience social isolation (Seckman, 2023). Maintaining social contact and mental engagement during lockdowns was associated with milder episodic memory impairments (Vakani et al., 2024). Moreover, individuals who reported a strong sense of belonging and greater participation in social activities also experienced better sleep quality (Xiao et al., 2020). Given that social interaction inherently involves cognitive stimulation, preserving strong social ties may serve as a protective factor for cognitive health, particularly during periods of crisis (Ingram et al., 2021).

To sum up, the cognitive effects of the COVID-19 pandemic are best understood as multifaceted and characterised by substantial individual variability. Moreover, these effects are highly context-dependent, shaped by the dynamic interplay between individual characteristics and the broader sociopolitical environment in which the crisis unfolded. Research has consistently shown that the impact of stress on cognitive functioning depends on multiple factors, including the intensity, timing, duration, and source of the stressor (James et al., 2023). During the pandemic, these factors varied considerably across countries and communities. Public health responses ranged from strict, prolonged lockdowns to more flexible, short-term restrictions—resulting in heterogeneous stressors and pandemic experiences. Consequently, the diverse cognitive outcomes observed across studies likely reflect not only methodological variability but also the cultural, political, and public health contexts that influenced how the pandemic was experienced worldwide.

5 Interventions to Alleviate Cognitive Consequences of the COVID-19 Pandemic

Even mild cognitive declines resulting from COVID-19 can significantly affect daily life, particularly for individuals with cognitively demanding tasks. Neurological and psychiatric symptoms can impair quality of life, independence, and social participation (Nolen et al., 2022; Tabacof et al., 2022; Yelin et al., 2022), yet many patients report difficulty accessing appropriate medical care (Espinoza & Martella, 2025).

Standard screening tools may miss cognitive dysfunction, highlighting the need for thorough cognitive assessment post-COVID (Crivelli et al., 2022). Persistent deficits call for targeted interventions, such as cognitive rehabilitation and person-centred care addressing sleep, fatigue, nutrition, and mental health (Espinoza & Martella, 2025; Poletti et al., 2022). Adapted attention and working memory training, psychoeducation, and compensatory strategies can be beneficial (Möller et al., 2023). Strong evidence supports cognitive training as an effective intervention for not only improving cognitive functioning, but also reducing PASC symptoms such as fatigue, sleep quality, mental health issues, and functional limitations (Weix et al., 2025). Some researchers recommend combining cognitive tasks with non-invasive brain stimulation, such as transcranial direct current stimulation, to enhance synaptic plasticity and modulate neural circuits underlying neurological, cognitive, and psychiatric disorders (Cavendish et al., 2022; Noda et al., 2023).

Patients with psychiatric comorbidities may benefit from combined psychotherapeutic, pharmacologic, and cognitive interventions, as over two-thirds report a combination of depression, fatigue, and cognitive complaints, predominantly memory problems and slowed processing speed (Farooqi et al., 2022). Based on a review of 24 trials, Zeraatkar et al. (2024) recommended cognitive–behavioural therapy—for which Weix et al. (2025) report moderate evidence—and physical/mental rehabilitation. However, some responses to Zeraatkar et al. (2024) criticised these interventions for potential harm (e.g., graded exercise may not always be safe) and for patient resistance to psychological framing. Patients often prefer their symptoms be treated as neurologically rather than psychologically grounded. It is essential to support them with mood management resources, cognitive compensation strategies, workplace/study accommodations, and education for both patients and their social circles (Krishnan et al., 2022; Möller et al., 2023).

Some studies have demonstrated the effectiveness of multidisciplinary interventions encompassing physical rehabilitation, cognitive training, and support for lifestyle modifications (Melillo et al., 2024; Weix et al., 2025). However, not all studies reported significant benefits, and further research is needed to establish the effectiveness of such multicomponent approaches (Melillo et al., 2024).

During crises such as the COVID-19 pandemic, individuals experiencing psychological distress and cognitive changes due to chronic stress may benefit from psychological and lifestyle interventions. These should promote sleep hygiene (Zhang et al., 2023), regular physical activity (Singh et al., 2023), cognitive engagement (Madeira et al., 2024), social interaction and resilience-building, as well as stress coping and social support (Gebhard et al., 2024). Additionally, access to mental healthcare and digital tools and remote interventions should be supported.

It is important to recognise that the direct and indirect effects of COVID-19 on cognition often result in similar symptom profiles—namely, deficits in attention, memory, executive function, and processing speed. Whether arising from neuroinvasive mechanisms and systemic inflammation in infected individuals or from chronic stress, social isolation, and disrupted routines in non-infected populations, the resulting cognitive outcomes were often clinically indistinguishable. Symptoms such as 'brain fog', fatigue, and executive dysfunction appeared across both groups, albeit with differing aetiologies and intensity (Table 1). This overlap highlights the need for integrated intervention frameworks that holistically address both cognitive and emotional health. Rather than treating viral and psychosocial consequences of the pandemic in isolation, effective care must target neurobiological and psychosocial dimensions in parallel. Interventions focusing solely on one aspect may prove insufficient; comprehensive approaches that combine neurorehabilitation with mental health support are essential for mitigating the long-term cognitive consequences of COVID-19.

6 Conclusion

Over time, the boundary between the direct neurological effects of SARS-CoV-2 infection and the indirect consequences of the pandemic, such as stress, social isolation, and uncertainty, has become increasingly difficult to disentangle. Nevertheless, converging evidence suggests that the pandemic impacted cognition through both pathways: directly, by affecting the central nervous system, and indirectly, through stress-related mechanisms that generated similar, albeit often less severe, cognitive symptoms. These outcomes were shaped by a complex interplay of individual vulnerabilities and contextual factors, resulting in diverse cognitive trajectories across the population. Crucially, symptoms like fatigue, executive dysfunction, and 'brain fog' were not limited to infected individuals, highlighting the broad neurocognitive toll of the pandemic.

Insights gained from this period, strengthened by ongoing cross-cultural and longitudinal studies, should inform future crisis responses. Decisions about imposing restrictive measures such as lockdowns must weigh not only infection control but also long-term cognitive and emotional health impacts. Mental health interventions in such contexts must explicitly support cognitive well-being, recognising the dynamic interdependence between emotional and cognitive functioning. Integrative care models that bring together neurologists, psychologists, occupational therapists, and other relevant health professionals are essential to addressing the multifaceted consequences of public health crises. Finally, public messaging should acknowledge that even subclinical cognitive complaints in the general population warrant attention, validation, and support—both to alleviate suffering and to prevent further deterioration.

Acknowledgement The authors acknowledge the financial support from the Slovenian Research Agency (research core funding No. P5-0110 and bilateral cooperation project No. BI-US/22-24-123).

References

Acharya, V., Fan, K., Snitz, B. E., Ganguli, M., DeKosky, S. T., Lopez, O. L., Feingold, E., & Kamboh, M. I. (2023). Meta-analysis of age-related cognitive decline reveals a novel locus for the attention domain and implicates a COVID-19-related gene for global cognitive function. *Alzheimer's & Dementia, 19*(11), 5010–5022. https://doi.org/10.1002/alz.13064

Al-Aly, Z., Xie, Y., & Bowe, B. (2021). High-dimensional characterization of post-acute sequelae of COVID-19. *Nature, 594*(7862), 259–264. https://doi.org/10.1038/s41586-021-03553-9

Alhola, P., & Polo-Kantola, P. (2007). Sleep deprivation: Impact on cognitive performance. *Neuropsychiatric Disease and Treatment, 3*(5), 553–567. https://doi.org/10.2147/ndt.s12160203

Almeria, M., Cejudo, J. C., Sotoca, J., Deus, J., & Krupinski, J. (2020). Cognitive profile following COVID-19 infection: Clinical predictors leading to neuropsychological impairment. *Brain, Behavior, & Immunity-Health, 9*, 100163. https://doi.org/10.1016/j.bbih.2020.100163

Altena, E., Micoulaud-Franchi, J. A., Geoffroy, P. A., Sanz-Arigita, E., Bioulac, S., & Philip, P. (2016). The bidirectional relation between emotional reactivity and sleep: From disruption to recovery. *Behavioral Neuroscience, 130*, 336–350. https://doi.org/10.1037/bne0000128

Amieva, H., Retuerto, N., Hernandez-Ruiz, V., Meillon, C., Dartigues, J. F., & Pérès, K. (2022). Longitudinal study of cognitive decline before and after the COVID-19 pandemic: Evidence from the PA-COVID survey. *Dementia and Geriatric Cognitive Disorders, 51*(1), 56–62. https://doi.org/10.1159/000521999

Ammar, A., Brach, M., Trabelsi, K., Chtourou, H., Boukhris, O., Masmoudi, L., Bouaziz, B., Bentlage, E., How, D., Ahmed, M., Müller, P., Müller, N., Aloui, A., Hammouda, O., Paineiras-Domingos, L., Braakman-Jansen, A., Wrede, C., Bastoni, S., et al. (2020a). Effects of COVID-19 home confinement on eating behaviour and physical activity: Results of the ECLB-COVID19 international online survey. *Nutrients, 12*(6), 1583. https://doi.org/10.3390/nu12061583

Ammar, A., Mueller, P., Trabelsi, K., Chtourou, H., Boukhris, O., Masmoudi, L., Bouaziz, B., Brach, M., Schmicker, M., Bentlage, E., How, D., Ahmed, M., Aloui, A., Hammouda, O., Paineiras-Domingos, L. L., Braakman-Jansen, A., Wrede, C., Bastoni, S., et al. (2020b). Psychological consequences of COVID-19 home confinement: The ECLB-COVID19 multicenter study. *PLoS One, 15*(11), e0240204. https://doi.org/10.1371/journal.pone.0240204

Ariza, M., Cano, N., Segura, B., Adan, A., Bargalló, N., Caldú, X., Campabadal, A., Jurado, M. A., Mataró, M., Pueyo, R., Sala-Llonch, R., Barrué, C., Bejar, J., Cortés, C. U., Junqué, C., & Garolera, M. (2022). Neuropsychological impairment in post-COVID condition individuals with and without cognitive complaints. *Frontiers in Aging Neuroscience, 14*, 1029842. https://doi.org/10.3389/fnagi.2022.1029842

Attanasi, G., Maffioletti, A., Shalukhina, T., Bel, C., & Cherikh, F. (2021). Gender differences in the impact of COVID-19 lockdown on potentially addictive behaviors: An emotion-mediated analysis. *Frontiers in Psychology, 12*, 703897. https://doi.org/10.3389/fpsyg.2021.703897

Austin, T. A., Thomas, M. L., Lu, M., Hodges, C. B., Darowski, E. S., Bergmans, R., Parr, S., Pickell, D., Catazaro, M., Lantrip, C., & Twamley, E. W. (2024). Meta-analysis of cognitive function following non-severe SARS-CoV-2 infection. *Neuropsychology Review*, 1–27. https://doi.org/10.1007/s11065-024-09642-6

Baglioni, C., Nanovska, S., Regen, W., Spiegelhalder, K., Feige, B., Nissen, C., Reynolds, C. F., III, & Riemann, D. (2016). Sleep and mental disorders: A meta-analysis of polysomnographic research. *Psychological Bulletin, 142*(9), 969–990. https://doi.org/10.1037/bul0000053

Batty, G. D., Deary, I. J., Luciano, M., Altschul, D. M., Kivimäki, M., & Gale, C. R. (2020). Psychosocial factors and hospitalisations for COVID-19: Prospective cohort study based on a community sample. *Brain, Behavior, and Immunity, 89*, 569–578. https://doi.org/10.1016/j.bbi.2020.06.02

Becker, J. H., Lin, J. J., Doernberg, M., Stone, K., Navis, A., Festa, J. R., & Wisnivesky, J. P. (2021). Assessment of cognitive function in patients after COVID-19 infection. *JAMA Network Open, 4*(10), e2130645. https://doi.org/10.1001/jamanetworkopen.2021.30645

Becker, J. H., Vannorsdall, T. D., & Weisenbach, S. L. (2023). Evaluation of Post–COVID-19 cognitive dysfunction: Recommendations for researchers. *JAMA Psychiatry, 80*(11), 1085–1086. https://doi.org/10.1001/jamapsychiatry.2023.2820

Borelli, W. V., Augustin, M. C., de Oliveira, P. B. F., Reggiani, L. C., Bandeira-de-Mello, R. G., Schumacher-Schuh, A. F., Chaves, M. L. F., & Castilhos, R. M. (2021). Neuropsychiatric symptoms in patients with dementia associated with increased psychological distress in caregivers during the COVID-19 pandemic. *Journal of Alzheimer's Disease, 80*(4), 1705–1712. https://doi.org/10.3233/jad-201513

Bourne Jr, L. E., & Yaroush, R. A. (2003). *Stress and cognition: A cognitive psychological perspective*. NASA/CR—2003-212282 Report (No. IH-045). https://ntrs.nasa.gov/api/citations/20040034070/downloads/20040034070.pdf

Brooks, S. K., Webster, R. K., Smith, L. E., Woodland, L., Wessely, S., Greenberg, N., & Rubin, G. J. (2020). The psychological impact of quarantine and how to reduce it: Rapid review of the evidence. *The Lancet, 395*(10227), 912–920. https://doi.org/10.1016/s0140-6736(20)30460-8

Bzdok, D., & Dunbar, R. I. (2020). The neurobiology of social distance. *Trends in Cognitive Sciences, 24*(9), 717–733. https://doi.org/10.1016/j.tics.2020.05.016

Carfì, A., Bernabei, R., & Landi, F. (2020). Persistent symptoms in patients after acute COVID-19. *JAMA, 324*(6), 603–605. https://doi.org/10.1001/jama.2020.12603

Caspersen, I. H., Magnus, P., & Trogstad, L. (2022). Excess risk and clusters of symptoms after COVID-19 in a large Norwegian cohort. *European Journal of Epidemiology, 37*(5), 539–548. https://doi.org/10.1007/s10654-022-00847-8

Cavaco, S., Sousa, G., Gonçalves, A., Dias, A., Andrade, C., Pereira, D., Aires, E. A., Moura, J., Silva, L., Varela, R., Malheiro, S., Oliveira, V., Teixeira-Pinto, A., Maia, L. F., & Correia, M. (2023). Predictors of cognitive dysfunction one-year post COVID-19. *Neuropsychology, 37*(5), 557–567. https://doi.org/10.1037/neu0000876

Cavendish, B. A., Lima, A., Bertola, L., Charvet, L., Bikson, M., Brunoni, A. R., & Vidal, K. S. (2022). Combination of transcranial direct current stimulation with online cognitive training improves symptoms of Post-acute Sequelae of COVID-19: A case series. *Brain Stimulation: Basic, Translational, and Clinical Research in Neuromodulation, 15*(6), 1375–1377. https://doi.org/10.1016/j.brs.2022.09.008

Ceban, F., Ling, S., Lui, L. M., Lee, Y., Gill, H., Teopiz, K. M., et al. (2022). Fatigue and cognitive impairment in Post-COVID-19 syndrome: A systematic review and meta-analysis. *Brain, Behavior, and Immunity, 101*, 93–135. https://doi.org/10.1016/j.bbi.2021.12.020

Cecchetti, G., Agosta, F., Canu, E., Basaia, S., Barbieri, A., Cardamone, R., Bernasconi, M. P., Castelnovo, V., Cividini, C., Cursi, M., Vabanesi, M., Impellizzeri, M., Lazzarin, S. M., Fanelli, G. F., Minicucci, F., Giacalone, G., Falini, A., Falautano, M., Rovere-Querini, P., et al. (2022). Cognitive, EEG, and MRI features of COVID-19 survivors: A 10-month study. *Journal of Neurology, 269*(7), 3400–3412. https://doi.org/10.1007/s00415-022-11047-5

Chattarji, S., Tomar, A., Suvrathan, A., Ghosh, S., & Rahman, M. M. (2015). Neighborhood matters: Divergent patterns of stress-induced plasticity across the brain. *Nature Neuroscience, 18*(10), 1364–1375. https://doi.org/10.1038/nn.4115

Chaumon, M., Rioux, P.-A., Herbst, S. K., Spiousas, I., Kübel, S. L., Gallego Hiroyasu, E. M., Runyun, Ş. L., Micillo, L., Thanopoulos, V., Mendoza-Duran, E., Wagelmans, A., Mudumba, R., Tachmatzidou, O., Cellini, N., D'Argembeau, A., Giersch, A., Grondin, S., Gronfier, C., Igarzábal, F. A., et al. (2022). The Blursday database as a resource to study subjective temporalities during COVID-19. *Nature Human Behaviour, 6*(11), 1587–1599. https://doi.org/10.1038/s41562-022-01419-2

Cohen, S. (2004). Social relationships and health. *American Psychologist, 59*(8), 676–684. https://doi.org/10.1037/0003-066X.59.8.676

Comijs, H. C., Deeg, D. J. H., Dik, M. G., Twisk, J. W. R., & Jonker, C. (2002). Memory complaints; the association with psycho-affective and health problems and the role of personality characteristics: A 6-year follow-up study. *Journal of Affective Disorders, 72*(2), 157–165. https://doi.org/10.1016/S0165-0327(01)00453-0

Corbett, A., Williams, G., Creese, B., Hampshire, A., Hayman, V., Palmer, A., Filakovszky, A., Mills, K., Cummings, J., Aarsland, D., Khan, Z., & Ballard, C. (2023). Cognitive decline in older adults in the UK during and after the COVID-19 pandemic: A longitudinal analysis of PROTECT study data. *The Lancet Healthy Longevity, 4*(11), e591–e599. https://doi.org/10.1016/s2666-7568(23)00187-3

Costas-Carrera, A., Sánchez-Rodríguez, M. M., Ojeda, A., Rodríguez-Rey, M. A., Martín-Villalba, I., Primé-Tous, M., Valdesoiro-Pulido, F., Segú, X., Borras, R., Clougher, D., Peri, J. M., & Vieta, E. (2024). Neuropsychological functioning and its correlates at 1 year follow-up of severe COVID-19. *Psychogeriatrics, 24*(4), 765–777. https://doi.org/10.1111/psyg.13113

Cousins, J. N., Sasmita, K., & Chee, M. W. (2018). Memory encoding is impaired after multiple nights of partial sleep restriction. *Journal of Sleep Research, 27*(1), 138–145. https://doi.org/10.1111/jsr.12578

Cristillo, V., Pilotto, A., Cotti Piccinelli, S., Zoppi, N., Bonzi, G., Gipponi, S., Sattin, D., Schiavolin, S., Raggi, A., Bezzi, M., Leonardi, M., & Padovani, A. (2021). Age and subtle cognitive impairment are associated with long-term olfactory dysfunction after COVID-19 infection. *Journal of the American Geriatrics Society, 69*(10), 2778–2780. https://doi.org/10.1111/jgs.17296

Crivelli, L., Palmer, K., Calandri, I., Guekht, A., Beghi, E., Carroll, W., Frontera, J., García-Azorín, D., Westenberg, E., Winkler, A. S., Mangialasche, F., Allegri, R. F., & Kivipelto, M. (2022). Changes in cognitive functioning after COVID-19: A systematic review and meta-analysis. *Alzheimer's & Dementia, 18*(5), 1047–1066. https://doi.org/10.1002/alz.12644

Davis, H. E., Assaf, G. S., McCorkell, L., Wei, H., Low, R. J., & Re'em, Y., Redfield, S., Austin, J. P., & Akrami, A. (2021). Characterizing long COVID in an international cohort: 7 months of symptoms and their impact. *EClinicalMedicine, 38*, 101019. https://doi.org/10.1016/j.eclinm.2021.101019

De Pue, S., Gillebert, C., Dierckx, E., Vanderhasselt, M. A., De Raedt, R., & Van den Bussche, E. (2021). The impact of the COVID-19 pandemic on wellbeing and cognitive functioning of older adults. *Scientific Reports, 11*(1), 4636. https://doi.org/10.1038/s41598-021-84127-7

De Tanti, A., Conforti, J., Bruni, S., De Gaetano, K., Cappalli, A., Basagni, B., Bertoni, D., & Saviola, D. (2023). Cognitive and psychological outcomes and follow-up in severely affected COVID-19 survivors admitted to a rehabilitation hospital. *Neurological Sciences, 44*(5), 1481–1489. https://doi.org/10.1007/s10072-023-06665-4

Delgado-Alonso, C., Valles-Salgado, M., Delgado-Álvarez, A., Yus, M., Gómez-Ruiz, N., Jorquera, M., Polidura, C., Gil, M. J., Marcos, A., Matías-Guiu, J., & Matías-Guiu, J. A. (2022). Cognitive dysfunction associated with COVID-19: A comprehensive neuropsychological study. *Journal of Psychiatric Research, 150*, 40–46. https://doi.org/10.1016/j.jpsychires.2022.03.033

Deng, J., Zhou, F., Hou, W., Silver, Z., Wong, C. Y., Chang, O., Huang, E., & Zuo, Q. K. (2021). The prevalence of depression, anxiety, and sleep disturbances in COVID-19 patients: A meta-analysis. *Annals of the New York Academy of Sciences, 1486*(1), 90–111. https://doi.org/10.1111/nyas.14506

Diamond, D. M., Campbell, A. M., Park, C. R., Halonen, J., & Zoladz, P. R. (2007). The temporal dynamics model of emotional memory processing: A synthesis on the neurobiological basis of stress-induced amnesia, flashbulb and traumatic memories, and the Yerkes-Dodson law. *Neural Plasticity, 2007*, 60803. https://doi.org/10.1155/2007/60803

Diaz, J., Herridge, M., Bertagnolio, S., Davis, H., Dua, T., Kaushic, C., Marshall, J., del Rosario Pérez, M., Strub-Wourgaft, N., & Soriano, J. (2021). Towards a universal understanding of post COVID-19 condition. *Bulletin of the World Health Organization, 99*(12), 901–903. https://doi.org/10.2471/BLT.21.286249

Dickerson, S. S., & Kemeny, M. E. (2004). Acute stressors and cortisol responses: A theoretical integration and synthesis of laboratory research. *Psychological Bulletin, 130*(3), 355–391. https://doi.org/10.1037/0033-2909.130.3.355

Díez-Cirarda, M., Yus-Fuertes, M., Sanchez-Sanchez, R., Gonzalez-Rosa, J. J., Gonzalez--Escamilla, G., Gil-Martínez, L., Delgado-Alonso, C., Gil-Moreno, M. J., Valles-Salgado, M., Cano-Cano, F., Ojeda-Hernandez, D., Gomez-Ruiz, N., Oliver-Mas, S., Benito-Martín, M. S., Jorquera, M., de la Fuente, S., Polidura, C., Selma-Calvo, B., Arrazola, J., et al. (2023). Hippocampal subfield abnormalities and biomarkers of pathologic brain changes: From SARS-CoV-2 acute infection to post-COVID syndrome. *EBioMedicine, 94*, 104711. https://doi.org/10.1016/j.ebiom.2023.104711

dos Santos, A. T., Leyendecker, D. D., Costa, A. L. S., & Souza-Talarico, J. N. D. (2012). Subjective memory complaint in healthy elderly: Influence of depressive symptoms, perceived stress and self-esteem. *Revista da Escola de Enfermagem da USP, 46*, 24–29. https://doi.org/10.1590/S0080-62342012000700004

Droit-Volet, S., Gil, S., Martinelli, N., Andant, N., Clinchamps, M., Parreira, L., Rouffiac, K., Dambrun, M., Huguet, P., Dubuis, B., Pereira, B., Bouillon, J.-B., & Dutheil, F. (2020). Time and Covid-19 stress in the lockdown situation: Time free, «Dying» of boredom and sadness. *PLoS One, 15*(8), e0236465. https://doi.org/10.1371/journal.pone.0236465

Espinoza, C., & Martella, D. (2025). Cognitive functions in COVID-19 survivors, approaches strategies, and impact on health systems: A qualitative systematic review. *European Archives of Psychiatry and Clinical Neuroscience, 275*(1), 5–49. https://doi.org/10.1007/s00406-023-01662-2

Fanshawe, J. B., Sargent, B. F., Badenoch, J. B., Saini, A., Watson, C. J., Pokrovskaya, A., Aniwattanapong, D., Conti, I., Nye, C., Burchill, E., Hussain, Z. U., Said, K., Kuhoga, E., Tharmaratnam, K., Pendered, S., Mbwele, B., Taquet, M., Wood, G. K., Rogers, J. P., Hampshire, A., et al. (2025). Cognitive domains affected post-COVID-19; a systematic review and meta-analysis. *European Journal of Neurology, 32*(1), e16181. https://doi.org/10.1111/ene.16181

Farooqi, M., Khan, A., Jacobs, A., D'Souza, V., Consiglio, F., Karmen, C. L., Dornbush, R., Hasnat, G. S., & Ferrando, S. J. (2022). Examining the long-term sequelae of SARS-CoV2 infection in patients seen in an outpatient psychiatric department. *Neuropsychiatric Disease and Treatment, 18*, 1259–1268. https://doi.org/10.2147/ndt.s357262

Farrer, L. M., Gulliver, A., Bennett, K., Fassnacht, D. B., & Griffiths, K. M. (2016). Demographic and psychosocial predictors of major depression and generalised anxiety disorder in Australian university students. *BMC Psychiatry, 16*, 241. https://doi.org/10.1186/s12888-016-0961-z

Favieri, F., Forte, G., Agostini, F., Giovannoli, J., Di Pace, E., Langher, V., Tambelli, R., Pazzaglia, M., Giannini, A., & Casagrande, M. (2021). The cognitive consequences of the COVID-19 pandemic on members of the general population in Italy: A preliminary study on executive inhibition. *Journal of. Clinical Medicine, 11*(1), 170. https://doi.org/10.3390/jcm11010170

Fazekas, C., Goswami, N., Matzer, F., Avian, A., Lodron, J., Rijksen, M., Hanfstingl, B., Kavcic, V., Groselj-Strele, A., Sourij, H., Kessler, H. H., Stelzl, E., Voegel, C. D., Binz, T. M., Schmid-Zalaudek, K., Wittmann, A., & Pilz, S. (2025). Perceived chronic stress prior to SARS-CoV-2 infection predicts ongoing symptomatic COVID-19: A prospective cohort study. *Psychotherapy and Psychosomatics*, 1–12. Portico. https://doi.org/10.1159/000547858

Fenollar-Cortés, J., Jiménez, Ó., Ruiz-García, A., & Resurrección, D. M. (2021). Gender differences in psychological impact of the confinement during the COVID-19 outbreak in Spain: A longitudinal study. *Frontiers in Psychology, 12*, 682860. https://doi.org/10.3389/fpsyg.2021.682860

Fernández Jiménez, E. M., Zavala-Calahorrano, A., & Aguilar Salazar, A. F. (2023). Effects of social isolation on the cognitive status of people over 65 years of age during the SARS-CoV-2 pandemic: A longitudinal comparative study. *Medwave, 23*(01), e2592. https://doi.org/10.5867/medwave.2023.01.2592

Ferrucci, R., Dini, M., Rosci, C., Capozza, A., Groppo, E., Reitano, M. R., Allocco, E., Poletti, B., Brugnera, A., Bai, F., Monti, A., Ticozzi, N., Silani, V., Centanni, S., D'Arminio Monforte, A., Tagliabue, L., & Priori, A. (2022). One-year cognitive follow-up of COVID-19 hospitalized patients. *European Journal of Neurology, 29*(7), 2006–2014. https://doi.org/10.1111/ene.15324

Fiorenzato, E., Zabberoni, S., Costa, A., & Cona, G. (2021). Cognitive and mental health changes and their vulnerability factors related to COVID-19 lockdown in Italy. *PLoS One, 16*(1), e0246204. https://doi.org/10.1371/journal.pone.0246204

Foreman, L., Child, B., Saywell, I., Collins-Praino, L., & Baetu, I. (2025). Cognitive reserve moderates the effect of COVID-19 on cognition: A systematic review and meta-analysis of individual participant data. *Neuroscience & Biobehavioral Reviews, 171*, 106067. https://doi.org/10.1016/j.neubiorev.2025.106067

Fu, W., Wang, C., Zou, L., et al. (2020). Psychological health, sleep quality, and coping styles to stress facing the COVID-19 in Wuhan, China. *Translational Psychiatry, 10*(1), 225. https://doi.org/10.1038/s41398-020-00913-3

Galasso, V., Pons, V., Profeta, P., Becher, M., Brouard, S., & Foucault, M. (2020). Gender differences in COVID-19 attitudes and behavior: Panel evidence from eight countries. *Proceedings of the National Academy of Sciences, 117*(44), 27285–27291. https://doi.org/10.1073/pnas.2012520117

Gebhard, C. E., Sütsch, C., Gebert, P., Gysi, B., Bengs, S., Todorov, A., Deforth, M., Buehler, P. K., Meisel, A., Schuepbach, R. A., Zinkernagel, A. S., Brugger, S. D., Acevedo, C., Patriki, D., Wiggli, B., Beer, J. H., Friedl, A., Twerenbold, R., Kuster, G. M., et al. (2024). Impact of sex and gender on post-COVID-19 syndrome, Switzerland, 2020. *Eurosurveillance, 29*(2), 2300200. https://doi.org/10.2807/1560-7917.es.2024.29.2.2300200

Giudice, V., Iannaccone, T., Faiella, F., Ferrara, F., Aversano, G., Coppola, S., De Chiara, E., Romano, M. G., Conti, V., & Filippelli, A. (2022). Gender differences in the impact of COVID-19 pandemic on mental health of Italian academic workers. *Journal of Personalized Medicine, 12*(4), 613. https://doi.org/10.3390/jpm12040613

Gómez-García, A. M., Arias Arias, A. J., Muñoz, F. L., & García-Rico, E. (2025). Allostatic load as a short-term prognostic and predictive marker. *Stress and Health, 41*(1), e3527. https://doi.org/10.1002/smi.3527

Gonzalez Aleman, G., Vavougios, G. D., Tartaglia, C., Uvais, N. A., Guekht, A., Hosseini, A. A., Lo Re, V., Ferreccio, C., D'Avossa, G., Zamponi, H. P., Figueredo Aguiar, M., Yecora, A., Ul Haq Katshu, M. Z., Stavrou, V. T., Boutlas, S., Gourgoulianis, K. I., Botero, C., González Insúa, F., Perez-Lloret, S., et al. (2025). Age-dependent phenotypes of cognitive impairment as sequelae of SARS-CoV-2 infection. *Frontiers in Aging Neuroscience, 16*, 1432357. https://doi.org/10.3389/fnagi.2024.1432357

Groff, D., Sun, A., Ssentongo, A. E., Ba, D. M., Parsons, N., Poudel, G. R., Lekoubou, A., Oh, J. S., Ericson, J. E., Ssentongo, P., & Chinchilli, V. M. (2021). Short-term and long-term rates of postacute sequelae of SARS-CoV-2 infection: A systematic review. *JAMA Network Open, 4*(10), e2128568. https://doi.org/10.1001/jamanetworkopen.2021.28568

Guardia, T., Cote, K. A., Healey, M. K., Gammage, K. L., & Campbell, K. L. (2024). Self-reported physical activity and sleep quality is associated with working memory function in middle-aged and older adults during the COVID-19 pandemic. *Aging, Neuropsychology, and Cognition, 31*(6), 1176–1195. https://doi.org/10.1080/13825585.2024.2333066

Hampshire, A., Trender, W., Chamberlain, S. R., Jolly, A. E., Grant, J. E., Patrick, F., Mazibuko, N., Williams, S. C., Barnby, J. M., Hellyer, P., & Mehta, M. A. (2021). Cognitive deficits in people who have recovered from COVID-19. *EClinicalMedicine, 39*, 101044. https://doi.org/10.1016/j.eclinm.2021.101044

Harb, A., da Silva, F. C., de Jesus, M. C. F., Willey, G., & Souza-Talarico, J. N. (2024). Loneliness during COVID-19: An overlooked risk factor for post-pandemic cognitive decline in older adults. *Alzheimer's & Dementia, 20*, e087817. https://doi.org/10.1002/alz.087817

Harlé, K. M., Shenoy, P., & in Paulus, M. P. (2013). The influence of emotions on cognitive control: Feelings and beliefs—where do they meet? *Frontiers in Human Neuroscience, 7*, 508. https://doi.org/10.3389/fnhum.2013.00508

Hausman, H. K., Dai, Y., O'Shea, A., Dominguez, V., Fillingim, M., Calfee, K., Carballo, D., Hernandez, C., Perryman, S., Kraft, J. N., Evangelista, N. D., Van Etten, E. J., Smith, S. G., Bharadwaj, P. K., Song, H., Porges, E., DeKosky, S. T., Hishaw, G. A., Marsiske, M., et al. (2022). The longitudinal impact of the COVID-19 pandemic on health behaviors, psychosocial factors, and cognitive functioning in older adults. *Frontiers in Aging Neuroscience, 14*, 999107. https://doi.org/10.3389/fnagi.2022.999107

Henckens, M. J., Hermans, E. J., Pu, Z., Joëls, M., & Fernández, G. (2009). Stressed memories: How acute stress affects memory formation in humans. *Journal of Neuroscience, 29*(32), 10111–10119. https://doi.org/10.1523/JNEUROSCI.1184-09.2009

Herrman, H., Stewart, D. E., Diaz-Granados, N., Berger, E. L., Jackson, B., & Yuen, T. (2011). What is resilience? *The Canadian Journal of Psychiatry, 56*(5), 258–265. https://doi.org/10.1177/070674371105600504

Holt-Lunstad, J., Smith, T. B., Baker, M., Harris, T., & Stephenson, D. (2015). Loneliness and social isolation as risk factors for mortality: A meta-analytic review. *Perspectives on Psychological Science, 10*(2), 227–237. https://doi.org/10.1177/1745691614568352

Hsu, W. C., Huang, N. C., & Hu, S. C. (2024). Gender differences in covid-19's effects on loneliness, isolation, and cognitive health among Taiwanese retirees. *Innovation in Aging, 8*(Suppl_1), 1003–1004. https://doi.org/10.1093/geroni/igae098.3231

Hwang, W. (2024). Transition patterns of intergenerational solidarity and digital communication during and after the COVID-19 pandemic in South Korea: Association with older parents' cognitive decline. *Applied Research in Quality of Life, 19*(3), 1271–1291. https://doi.org/10.1007/s11482-024-10284-2

Ingram, J., Hand, C. J., & Maciejewski, G. (2021). Social isolation during COVID-19 lockdown impairs cognitive function. *Applied Cognitive Psychology, 35*(4), 935–947. https://doi.org/10.1002/acp.3821

Jacob, L., Haro, J. M., & Koyanagi, A. (2019). Physical multimorbidity and subjective cognitive complaints among adults in the United Kingdom: A cross-sectional community-based study. *Scientific Reports, 9*(1), 12417. https://doi.org/10.1038/s41598-019-48894-8

Jahrami, H., BaHammam, A. S., Bragazzi, N. L., Saif, Z., Faris, M., & Vitiello, M. V. (2021). Sleep problems during the COVID-19 pandemic by population: A systematic review and meta-analysis. *Journal of Clinical Sleep Medicine, 17*(2), 299–313. https://doi.org/10.5664/jcsm.8930

James, K. A., Stromin, J. I., Steenkamp, N., & Combrinck, M. I. (2023). Understanding the relationships between physiological and psychosocial stress, cortisol and cognition. *Frontiers in Endocrinology, 14*, 1085950. https://doi.org/10.3389/fendo.2023.1085950

Jiang, Y., Neal, J., Sompol, P., Yener, G., Arakaki, X., Norris, C. M., Farina, F. R., Ibanez, A., Lopez, S., Al-Ezzi, A., Kavcic, V., Güntekin, B., Babiloni, C., & Hajós, M. (2024). Parallel electrophysiological abnormalities due to COVID-19 infection and to Alzheimer's disease and related dementia. *Alzheimer's & Dementia, 20*(10), 7296–7319. https://doi.org/10.1002/alz.14089

Johnston, S., Brenu, E. W., Staines, D., & Marshall-Gradisnik, S. (2013). The prevalence of chronic fatigue syndrome/ myalgic encephalomyelitis: a meta-analysis. *Clinical Epidemiology, 5*, 105–110. https://doi.org/10.2147/CLEP.S39876

Kemeny, M. E. (2003). The psychobiology of stress. *Current Directions in Psychological Science, 12*(4), 124–129. https://doi.org/10.1111/1467-8721.01246

Knapp, S. A. B., Austin, D. S., Aita, S. L., Caron, J. E., Owen, T., Borgogna, N. C., Del Bene, V. A., Roth, R. M., Milberg, W. P., & Hill, B. D. (2024). Neurocognitive and psychiatric outcomes associated with postacute COVID-19 infection without severe medical complication: A meta-analysis. *Journal of Neurology, Neurosurgery & Psychiatry, 95*(12), 1207–1216. https://doi.org/10.1136/jnnp-2024-333950

Kniffin, K. M., Narayanan, J., Anseel, F., Antonakis, J., Ashford, S. P., Bakker, A. B., Bamberger, P., Bapuji, H., Bhave, D. P., Choi, V. K., Creary, S. J., Demerouti, E., Flynn, F. J., Gelfand, M. J., Greer, L. L., Johns, G., Kesebir, S., Klein, P. G., Lee, S. Y., et al. (2021). COVID-19 and the workplace: Implications, issues, and insights for future research and action. *American Psychologist, 76*(1), 63–77. https://doi.org/10.1037/amp0000716

Kocalevent, R. D., Hinz, A., Brähler, E., & Klapp, B. F. (2011). Determinants of fatigue and stress. *BMC Research Notes, 4*, 238. https://doi.org/10.1186/1756-0500-4-238

Krishnamoorthy, Y., Nagarajan, R., Saya, G. K., & Menon, V. (2020). Prevalence of psychological morbidities among general population, healthcare workers and COVID-19 patients amidst the COVID-19 pandemic: A systematic review and meta-analysis. *Psychiatry Research, 293*, 113382. https://doi.org/10.1016/j.psychres.2020.113382

Krishnan, K., Miller, A. K., Reiter, K., & Bonner-Jackson, A. (2022). Neurocognitive profiles in patients with persisting cognitive symptoms associated with COVID-19. *Archives of Clinical Neuropsychology, 37*(4), 729–737. https://doi.org/10.1093/arclin/acac004

Krueger, G. P. (1989). Sustained work, fatigue, sleep loss and performance: A review of the issues. *Work & Stress, 3*(2), 129–141. https://doi.org/10.1080/02678378908256939

Leng, Y., Wainwright, N. W. J., Hayat, S., Stephan, B. C. M., Matthews, F. E., Luben, R., Surtees, P. G., Khaw, K.-T., & Brayne, C. (2013). The association between social stress and global cognitive function in a population-based study: The European Prospective Investigation into Cancer (EPIC)-Norfolk study. *Psychological Medicine, 43*(3), 655–666. https://doi.org/10.1017/S0033291712001316

Li, Y., Godai, K., Kido, M., Komori, S., Shima, R., Kamide, K., & Kabayama, M. (2022). Cognitive decline and poor social relationship in older adults during COVID-19 pandemic: Can information and communications technology (ICT) use helps? *BMC Geriatrics, 22*(1), 375. https://doi.org/10.1186/s12877-022-03061-z

Li, Y., Wang, A., Wu, Y., Han, N., & Huang, H. (2021). Impact of the COVID-19 pandemic on the mental health of college students: A systematic review and meta-analysis. *Frontiers in Psychology, 12*, 669119. https://doi.org/10.3389/fpsyg.2021.669119

Li, Z., Zhang, Z., Zhang, Z., Wang, Z., & Li, H. (2023). Cognitive impairment after long COVID-19: Current evidence and perspectives. *Frontiers in Neurology, 14*, 1239182. https://doi.org/10.3389/fneur.2023.1239182

Liston, C., McEwen, B. S., & Casey, B. J. (2009). Psychosocial stress reversibly disrupts prefrontal processing and attentional control. *Proceedings of the National Academy of Sciences, 106*(3), 912–917. https://doi.org/10.1073/pnas.0807041106

Liu, C. H., Zhang, E., Wong, G. T. F., Hyun, S., & Hahm, H. C. (2020). Factors associated with depression, anxiety, and PTSD symptomatology during the COVID-19 pandemic: Clinical implications for U.S. young adult mental health. *Psychiatry Research, 290*, 113172. https://doi.org/10.1016/j.psychres.2020.113172

Liu, Y.-H., Chen, Y., Wang, Q.-H., Wang, L.-R., Jiang, L., Yang, Y., Chen, X., Li, Y., Cen, Y., Xu, C., Zhu, J., Li, W., Wang, Y.-R., Zhang, L.-L., Liu, J., Xu, Z.-Q., & Wang, Y.-J. (2022). One-year trajectory of cognitive changes in older survivors of COVID-19 in Wuhan, China: A longitudinal cohort study. *JAMA Neurology, 79*(5), 509–517. https://doi.org/10.1001/jamaneurol.2022.0461

Llana, T., Mendez, M., Garces-Arilla, S., Hidalgo, V., Mendez-Lopez, M., & Juan, M. C. (2023). Association between olfactory dysfunction and mood disturbances with objective and subjective cognitive deficits in long-COVID. *Frontiers in Psychology, 14*, 1076743. https://doi.org/10.3389/fpsyg.2023.1076743

Llana, T., Zorzo, C., Mendez-Lopez, M., & Mendez, M. (2024). Memory alterations after COVID-19 infection: A systematic review. *Applied Neuropsychology: Adult, 31*(3), 292–305. https://doi.org/10.1080/23279095.2022.2123739

Logue, J. K., Franko, N. M., McCulloch, D. J., McDonald, D., Magedson, A., Wolf, C. R., & Chu, H. Y. (2021). Sequelae in adults at 6 months after COVID-19 infection. *JAMA Network Open, 4*(2), e210830. https://doi.org/10.1001/jamanetworkopen.2021.0830

Madeira, E. S., de Souza, P. A., & Amaral, A. (2024). Remote activities of cognitive stimulation for older adults during the COVID-19 pandemic: A systematic review. *Cadernos de Saúde Pública, 40*(2), e00081923. https://doi.org/10.1590/0102-311xen081923

Maggi, G., Baldassarre, I., Barbaro, A., Cavallo, N. D., Cropano, M., Nappo, R., & Santangelo, G. (2022). Age-and gender-related differences in the evolution of psychological and cognitive status after the lockdown for the COVID-19 outbreak: A follow-up study. *Neurological Sciences, 43*(3), 1521–1532. https://doi.org/10.1007/s10072-021-05768-0

Mazza, M. G., Palladini, M., De Lorenzo, R., Magnaghi, C., Poletti, S., Furlan, R., Ciceri, F., Rovere--Querini, P., & Benedetti, F. (2021). Persistent psychopathology and neurocognitive impairment in COVID-19 survivors: Effect of inflammatory biomarkers at three-month follow-up. *Brain, Behavior, and Immunity, 94*, 138–147. https://doi.org/10.1016/j.jpsychires.2022.08.008

McGaugh, J. L. (2015). Consolidating memories. *Annual Review of Psychology, 66*(1), 1–24. https://doi.org/10.1146/annurev-psych-010814-014954

McWhirter, L., Smyth, H., Hoeritzauer, I., Couturier, A., Stone, J., & Carson, A. J. (2023). What is brain fog? *Journal of Neurology, Neurosurgery & Psychiatry, 94*(4), 321–325. https://doi.org/10.1136/jnnp-2022-329683

Mekhora, C., Lamport, D. J., & Spencer, J. P. E. (2024). An overview of the relationship between inflammation and cognitive function in humans, molecular pathways and the impact of nutraceuticals. *Neurochemistry International, 181*, 105900. https://doi.org/10.1016/j.neuint.2024.105900

Melillo, A., Perrottelli, A., Caporusso, E., Coltorti, A., Giordano, G. M., Giuliani, L., Pezzella, P., Bucci, P., Mucci, A., Galderisi, S., & Maj, M. (2024). Research evidence on the management of the cognitive impairment component of the post-COVID condition: A qualitative systematic review. *European Psychiatry, 67*(1), e60. https://doi.org/10.1192/j.eurpsy.2024.1770

Méndez, R., Balanzá-Martínez, V., Luperdi, S. C., Estrada, I., Latorre, A., González-Jiménez, P., Feced, L., Bouzas, L., Yépez, K., Ferrando, A., Hervás, D., Zaldívar, E., Reyes, S., Berk, M., & Menéndez, R. (2021). Short-term neuropsychiatric outcomes and quality of life in COVID-19 survivors. *Journal of Internal Medicine, 290*(3), 621–631. https://doi.org/10.1111/joim.13262

Menze, I., Mueller, P., Mueller, N. G., & Schmicker, M. (2022). Age-related cognitive effects of the COVID-19 pandemic restrictions and associated mental health changes in Germans. *Scientific Reports, 12*(1), 8172. https://doi.org/10.1038/s41598-022-11283-9

Milde, C., Glombiewski, J. A., Wilhelm, M., & Schemer, L. (2023). Psychological factors predict higher odds and impairment of post-COVID symptoms: A prospective study. *Psychosomatic Medicine, 85*(6), 479–487. https://doi.org/10.1097/psy.0000000000001214

Miller, L. R., Divers, R., Reed, C., Pugh, E., & Calamia, M. (2021). Resilience as a moderator of depression and anxiety: A bidimensional approach to predictors of subjective cognition in older adults. *Aging & Mental Health, 27*(1), 29–34. https://doi.org/10.1080/13607863.2021.2013432

Miskowiak, K. W., Pedersen, J. K., Gunnarsson, D. V., Roikjer, T. K., Podlekareva, D., Hansen, H., Dall, C. H., & Johnsen, S. (2023). Cognitive impairments among patients in a long-COVID clinic: Prevalence, pattern and relation to illness severity, work function and quality of life. *Journal of Affective Disorders, 324*, 162–169. https://doi.org/10.1016/j.jad.2022.12.122

Miskowiak, K., Johnsen, S., Sattler, S., Nielsen, S., Kunalan, K., Rungby, J., Lapperre, T., & Porsberg, C. (2021). Cognitive impairments four months after COVID-19 hospital discharge: Pattern, severity and association with illness variables. *European Neuropsychopharmacology, 46*, 39–48. https://doi.org/10.1016/j.euroneuro.2021.03.019

Möller, M., Borg, K., Janson, C., Lerm, M., Normark, J., & Niward, K. (2023). Cognitive dysfunction in post-COVID-19 condition: Mechanisms, management, and rehabilitation. *Journal of Internal Medicine, 294*(5), 563–581. https://doi.org/10.1111/joim.13720

Montoliu, T., Hidalgo, V., Pulopulos, M. M., Ivorra, J. L., Martínez, M. J., & Salvador, A. (2018). The relationship between cortisol and cognitive function in healthy older people: The moderating role of Apolipoprotein E polymorphism. *Neurobiology of Learning and Memory, 155*, 297–305. https://doi.org/10.1016/j.nlm.2018.08.013

Nair, N., Hegarty, J. P., II, Ferguson, B. J., Hecht, P. M., Tilley, M., Christ, S. E., & Beversdorf, D. Q. (2020a). Effects of stress on functional connectivity during problem solving. *NeuroImage, 208*, 116407. https://doi.org/10.1016/j.neuroimage.2019.116407

Nair, N., Hegarty, J. P., Ferguson, B. J., Hooshmand, S. J., Hecht, P. M., Tilley, M., Christ, S. E., & Beversdorf, D. Q. (2020b). Effects of stress on functional connectivity during verbal processing. *Brain Imaging and Behavior, 14*, 2708–2723. https://doi.org/10.1007/s11682-019-00221-5

Noda, Y., Sato, A., Shichi, M., Sato, A., Fujii, K., Iwasa, M., Nagano, Y., Kitahata, R., & Osawa, R. (2023). Real world research on transcranial magnetic stimulation treatment strategies for neuropsychiatric symptoms with long-COVID in Japan. *Asian Journal of Psychiatry, 81*, 103438. https://doi.org/10.1016/j.ajp.2022.103438

Nogueira, J., Gerardo, B., Silva, A. R., Pinto, P., Barbosa, R., Soares, S., Baptista, B., Paquete, C., Cabral-Pinto, M., Vilar, M. M., Simões, M. R., & Freitas, S. (2021). Effects of restraining measures due to COVID-19: Pre- and post-lockdown cognitive status and mental health. *Current Psychology, 41*(10), 7383–7392. https://doi.org/10.1007/s12144-021-01747-y

Nolen, L. T., Mukerji, S. S., & Mejia, N. I. (2022). Post-acute neurological consequences of COVID-19: An unequal burden. *Nature Medicine, 28*(1), 20–23. https://doi.org/10.1038/s41591-021-01647-5

Novais, A., Monteiro, S., Roque, S., Correia-Neves, M., & Sousa, N. (2017). How age, sex and genotype shape the stress response. *Neurobiology of Stress, 6*, 44–56. https://doi.org/10.1016/j.ynstr.2016.11.004

Oyejide, A. O., Besharati, S. N., Alcock, S., Schioth, H. B., & Brooks, S. J. (2025). A global survey on the associations between the lockdown group, free memory recall and emotional responses during the COVID-19 lockdown. *Scientific Reports, 15*, 12220. https://doi.org/10.1038/s41598-025-91991-0

Paolini, M., Palladini, M., Mazza, M. G., Colombo, F., Vai, B., Rovere-Querini, P., Falini, A., Poletti, S., & Benedetti, F. (2023). Brain correlates of subjective cognitive complaints in COVID-19 survivors: A multimodal magnetic resonance imaging study. *European Neuropsychopharmacology, 68*, 1–10. https://doi.org/10.1016/j.euroneuro.2022.12.002

Pieh, C., Budimir, S., Delgadillo, J., Barkham, M., Fontaine, J. R., & Probst, T. (2021). Mental health during COVID-19 lockdown in the United Kingdom. *Biopsychosocial Science and Medicine, 83*(4), 328–337. https://doi.org/10.1097/psy.0000000000000871

Podlesek, A., & Kavcic, V. (2021). Generalised anxiety in Slovenian university students during the COVID-19 pandemic. *Journal of Contemporary Educational Studies, 72*(138), 322–341.

Podlesek, A., Komidar, L., & Kavcic, V. (2021). The relationship between perceived stress and subjective cognitive decline during the COVID-19 epidemic. *Frontiers in Psychology, 12*, 647971. https://doi.org/10.3389/fpsyg.2021.647971

Poletti, S., Palladini, M., Mazza, M. G., De Lorenzo, R., Irene, B., Sara, B., Beatrice, B., Ceciclio, B., Stefania, C., Valentina, C., Elisa, C., Jacopo, C., Marta, C., Elena, C., Federica, C., Sarah, D., Greta, D., Camilla, D. P., et al. (2022). Long-term consequences of COVID-19 on cognitive functioning up to 6 months after discharge: Role of depression and impact on quality of life. *European Archives of Psychiatry and Clinical Neuroscience, 272*, 773–782. https://doi.org/10.1007/s00406-021-01346-9

Porcelli, A. J., & Delgado, M. R. (2017). Stress and decision making: Effects on valuation, learning, and risk-taking. *Current Opinion in Behavioral Sciences, 14*, 33–39. https://doi.org/10.1016/j.cobeha.2016.11.015

Premraj, L., Kannapadi, N. V., Briggs, J., Seal, S. M., Battaglini, D., Fanning, J., Suen, J., Robba, C., Fraser, J., & Cho, S.-M. (2022). Mid and long-term neurological and neuropsychiatric manifestations of post-COVID-19 syndrome: A meta-analysis. *Journal of the Neurological Sciences, 434*, 120162. https://doi.org/10.1016/j.jns.2022.120162

Qin, S., Hermans, E. J., van Marle, H. J. F., Luo, J., & Fernández, G. (2009). Acute psychological stress reduces working memory-related activity in the dorsolateral prefrontal cortex. *Biological Psychiatry, 66*(1), 25–32. https://doi.org/10.1016/j.biopsych.2009.03.006

Rajan, S., Khunti, K., Alwan, N., Steves, C., Greenhalgh, T., MacDermott, N., Sagan, A., & McKee, M. (2021). *In the wake of the pandemic. Preparing for long COVID*. European Observatory on Health Systems and Policies. https://apps.who.int/iris/bitstream/handle/10665/339629/Policy--brief-39-1997-8073-eng.pdf

Reddy, R., Palmer, C. A., Jackson, C., Farris, S. G., & Alfano, C. A. (2017). Impact of sleep restriction versus idealized sleep on emotional experience, reactivity and regulation in healthy adolescents. *Journal of Sleep Research, 26*, 516e25. https://doi.org/10.1111/jsr.12484

Reisch, T., Heiler, G., Hurt, J., Klimek, P., Hanbury, A., & Thurner, S. (2021). Behavioral gender differences are reinforced during the COVID-19 crisis. *Scientific Reports, 11*(1), 19241. https://doi.org/10.1038/s41598-021-97394-1

Roozendaal, B., McEwen, B. S., & Chattarji, S. (2009). Stress, memory and the amygdala. *Nature Reviews Neuroscience, 10*, 423–433. https://doi.org/10.1038/nrn2651

Sandi, C. (2013). Stress and cognition. *WIREs Cognitive Science, 4*(3), 245–261. https://doi.org/10.1002/wcs.1222

Sandi, C., & Haller, J. (2015). Stress and the social brain: Behavioural effects and neurobiological mechanisms. *Nature Reviews Neuroscience, 16*, 290–304. https://doi.org/10.1038/nrn3918

Santangelo, G., Baldassarre, I., Barbaro, A., Cavallo, N. D., Cropano, M., Maggi, G., Nappo, R., Trojano, L., & Raimo, S. (2021). Subjective cognitive failures and their psychological correlates in a large Italian sample during quarantine/self-isolation for COVID-19. *Neurological Sciences, 42*(7), 2625–2635. https://doi.org/10.1007/s10072-021-05268-1

Schneiderman, N., Ironson, G., & Siegel, S. D. (2005). Stress and health: Psychological, behavioral, and biological determinants. *Annual Review of Clinical Psychology, 1*, 607–528. https://doi.org/10.1146/annurev.clinpsy.1.102803.144141

Schou, T. M., Joca, S., Wegener, G., & Bay-Richter, C. (2021). Psychiatric and neuropsychiatric sequelae of COVID-19–A systematic review. *Brain, Behavior, and Immunity, 97*, 328–348. https://doi.org/10.1016/j.bbi.2021.07.018

Schrempft, S., Baysson, H., Graindorge, C., Pullen, N., Hagose, M., Zaballa, M.-E., Preisig, M., Nehme, M., Guessous, I., Stringhini, S., Arm-Vernez, I., Azman, A. S., Bachmann, D., Bal, A., Balavoine, J.-F., Barbe, R. P., Baysson, H., Berthelot, J., Bouhet, A. R., et al. (2024). Biopsychosocial risk factors for subjective cognitive decline among older adults during the COVID-19 pandemic: A population-based study. *Public Health, 234*, 16–23. https://doi.org/10.1016/j.puhe.2024.05.025

Schwabe, L., & Wolf, O. T. (2013). Stress and multiple memory systems: From 'thinking' to 'doing'. *Trends in Cognitive Sciences, 17*(2), 60–68. https://doi.org/10.1016/j.tics.2012.12.001

Schwabe, L., Joëls, M., Roozendaal, B., Wolf, O. T., & Oitzl, M. S. (2012). Stress effects on memory: An update and integration. *Neuroscience & Biobehavioral Reviews, 36*(7), 1740–1749. https://doi.org/10.1016/j.neubiorev.2011.07.002

Scullin, M. K., & Bliwise, D. L. (2015). Sleep, cognition, and normal aging: Integrating a half century of multidisciplinary research. *Perspectives on Psychological Science, 10*(1), 97–137. https://doi.org/10.1177/1745691614556680

Seckman, C. (2023). The impact of COVID-19 on the psychosocial well-being of older adults: A literature review. *Journal of Nursing Scholarship, 55*(1), 97–111.

Selye, H. (1956). *The stress of life*. McGraw-Hill.

Shields, G. S., Sazma, M. A., & Yonelinas, A. P. (2016). The effects of acute stress on core executive functions: A meta-analysis and comparison with cortisol. *Neuroscience & Biobehavioral Reviews, 68*, 651–668. https://doi.org/10.1016/j.neubiorev.2016.06.038

Silva, T. C. D., Scarmagnan, G. S., Batiston, A. P., Santos, M. L. D. M. D., & Christofoletti, G. (2022). Impact of the COVID-19 pandemic on the cognitive and motor functions of older people: A 3-year cohort study. *Revista Brasileira de Geriatria e. Gerontologia, 25*, e220146. https://doi.org/10.1590/1981-22562022025.220146.en

Singh, B., Olds, T., Curtis, R., Dumuid, D., Virgara, R., Watson, A., Szeto, K., O'Connor, E., Ferguson, T., Eglitis, E., Miatke, A., Simpson, C. E., & Maher, C. (2023). Effectiveness of physical activity interventions for improving depression, anxiety and distress: An overview of systematic reviews. *British Journal of Sports Medicine, 57*(18), 1203–1209. https://doi.org/10.1136/bjsports-2022-106195

Soriano, J. B., Murthy, S., Marshall, J. C., Relan, P., & Diaz, J. V. (2022). A clinical case definition of post-COVID-19 condition by a Delphi consensus. *The Lancet Infectious Diseases, 22*(4), e102–e107. https://doi.org/10.1016/s1473-3099(21)00703-9

Steinberg, S. I., Negash, S., Sammel, M. D., Bogner, H., Harel, B. T., Livney, M. G., McCoubrey, H., Wolk, D. A., Kling, M. A., & Arnold, S. E. (2013). Subjective memory complaints, cognitive performance, and psychological factors in healthy older adults. *American Journal of Alzheimer's Disease & Other Dementias®, 28*(8), 776–783. https://doi.org/10.1177/1533317513504817

Stenfors, C. U., Marklund, P., Magnusson Hanson, L. L., Theorell, T., & Nilsson, L. G. (2013). Subjective cognitive complaints and the role of executive cognitive functioning in the working population: A case-control study. *PLoS One, 8*(12), e83351. https://doi.org/10.1371/journal.pone.0083351

Stern, Y., Arenaza-Urquijo, E. M., Bartrés-Faz, D., Belleville, S., Cantilon, M., Chetelat, G., Ewers, M., Franzmeier, N., Kempermann, G., Kremen, W. S., Okonkwo, O., Scarmeas, N., Soldan, A., Udeh-Momoh, C., Valenzuela, M., Vemuri, P., & Vuoksimaa, E. (2020). Whitepaper: Defining and investigating cognitive reserve, brain reserve, and brain maintenance. *Alzheimer's Dementia, 16*(9), 1305–1311. https://doi.org/10.1016/j.jalz.2018.07.219

Tabacof, L., Tosto-Mancuso, J., Wood, J., Cortes, M., Kontorovich, A., McCarthy, D., Rizk, D., Mohammadi, N., Breyman, E., Nasr, L., Kellner, C., & Putrino, D. (2022). Post-acute COVID-19 syndrome negatively impacts physical function, cognitive function, health-related quality of life, and participation. *American Journal of Physical Medicine & Rehabilitation, 101*(1), 48–52. https://doi.org/10.1097/PHM.0000000000001910

Tanaka, S., Saida, K., Murayama, A., Higuchi, D., & Shinohara, T. (2023). Associated factors of new subjective cognitive decline complaints after a 6-month period among community-dwelling older adults during the COVID-19 pandemic in Japan. *Psychogeriatrics, 23*(1), 136–140. https://doi.org/10.1111/psyg.12917

Taquet, M., Sillett, R., Zhu, L., Mendel, J., Camplisson, I., Dercon, Q., & Harrison, P. J. (2022). Neurological and psychiatric risk trajectories after SARS-CoV-2 infection: An analysis of 2-year retrospective cohort studies including 1 284 437 patients. *The Lancet Psychiatry, 9*(10), 815–827. https://doi.org/10.1016/s2215-0366(22)00260-7

Tavares-Júnior, J. W. L., Oliveira, D. N., da Silva, J. B. S., Queiroz Feitosa, W. L., Sousa, A. V. M., Marinho, S. C., Cunha, L. C. V., de Gaspar, S., Gomes, C. M. P., de Oliveira, L. L. B., Moreira--Nunes, C. A., Sobreira, E. S. T., de Moraes, M. E. A., Sobreira-Neto, M. A., Montenegro, R. C., & Braga-Neto, P. (2023). Post-COVID-19 cognitive decline and apoe polymorphism: Towards a possible link? *Brain Sciences, 13*(12), 1611. https://doi.org/10.3390/brainsci13121611

Taylor, A. H., & Dorn, L. (2006). Stress, fatigue, health, and risk of road traffic accidents among professional drivers: The contribution of physical inactivity. *Annual Review of Public Health, 27*(1), 371–391. https://doi.org/10.1146/annurev.publhealth.27.021405.102117

Thorsteinsson, E. B., Brown, R. F., & Owens, M. T. (2019). Modeling the effects of stress, anxiety, and depression on rumination, sleep, and fatigue in a nonclinical sample. *The Journal of Nervous and Mental Disease, 207*(5), 355–359. https://doi.org/10.1097/NMD.0000000000000973

Uchino, B. N. (2009). Understanding the links between social support and physical health: A life-span perspective with emphasis on the separability of perceived and received support. *Perspectives on Psychological Science, 4*(3), 236–255. https://doi.org/10.1111/j.1745-6924.2009.01122.x

Vakani, K., Ratto, M., Sandford-James, A., Antonova, E., & Kumari, V. (2024). Cognitive and mental health trajectories of COVID-19: Role of hospitalisation and long-COVID symptoms. *European Psychiatry, 67*(1), e17. https://doi.org/10.1192/j.eurpsy.2024.7

Valerio, T. D., Kim, J., Sexton-Radek, K., & Kim, M. J. (2016). Association of stress, general health, and alcohol use with poor sleep quality among U.S. college students. *American Journal of Health Education, 47*(1), 17–23. https://doi.org/10.1080/19325037.2015.1111173

Valla, L. G., Rossi, M., Gaia, A., Guaita, A., & Rolandi, E. (2025). The impact of the COVID-19 pandemic on oldest-old social capital and health and the role of digital inequalities: Longitudinal cohort study. *Journal of Medical Internet Research, 27*, e62824. https://doi.org/10.2196/62824

van der Feltz-Cornelis, C., Turk, F., Sweetman, J., Khunti, K., Gabbay, M., Shepherd, J., Montgomery, H., Strain, W. D., Lip, G. Y. H., Wootton, D., Watkins, C. L., Cuthbertson, D. J., Williams, N., & Banerjee, A. (2024). Prevalence of mental health conditions and brain fog in people with long COVID: A systematic review and meta-analysis. *General Hospital Psychiatry, 88*, 10–22. https://doi.org/10.1016/j.genhosppsych.2024.02.009

Van Reeth, O., Weibel, L., Spiegel, K., Leproult, R., Dugovic, C., & Maccari, S. (2000). Physiology of sleep (review)–interactions between stress and sleep: From basic research to clinical situations. *Sleep Medicine Reviews, 4*(2), 201–219. https://doi.org/10.1053/smrv.1999.0097

Vance, D. E., Fazeli, P. L., & Lokken, K. (2025). New reflections on COVID-19's implications for neurocognition. *JAMA Network Open, 8*(4), e255539. https://doi.org/10.1001/jamanetworkopen.2025.5539

Varga, Z., Flammer, A. J., Steiger, P., Haberecker, M., Andermatt, R., Zinkernagel, A. S., Mehra, M. R., Schuepbach, R. A., Ruschitzka, F., & Moch, H. (2020). Endothelial cell infection and endotheliitis in COVID-19. *The Lancet, 395*(10234), 1417–1418. https://doi.org/10.1016/s0140-6736(20)30937-5

Wang, S., Menor, A., Chibnik, L. B., Kang, J. H., Vyas, C. M., Blacker, D. L., Kubzansky, L. D., Koenen, K. C., & Roberts, A. L. (2025). COVID-19 pandemic–related exposures and cognitive function in middle-aged women. *JAMA Network Open, 8*(4), e255532. https://doi.org/10.1001/jamanetworkopen.2025.5532

Wang, S., Quan, L., Chavarro, J. E., Slopen, N., Kubzansky, L. D., Koenen, K. C., Kang, J. H., Weisskopf, M. G., Branch-Elliman, W., & Roberts, A. L. (2022). Associations of depression, anxiety, worry, perceived stress, and loneliness prior to infection with risk of post–COVID-19 conditions. *JAMA Psychiatry, 79*(11), 1081–1091. https://doi.org/10.1001/jamapsychiatry.2022.2640

Warren, S. L., Reid, E., Whitfield, P., & Moustafa, A. A. (2022). Subjective memory complaints as a predictor of mild cognitive impairment and Alzheimer's disease. *Discover Psychology, 2*(1), 13. https://doi.org/10.1007/s44202-022-00031-9

Weix, N. M., Shake, H. M., Duran Saavedra, A. F., Clingan, H. E., Hernandez, V. C., Johnson, G. M., Hansen, A. D., Collins, D. M., Pryor, L. E., Kitchens, R., Armstead, A., & Hilton, C. (2025). Cognitive interventions and rehabilitation to address Long-COVID symptoms: A systematic review. *OTJR: Occupational Therapy Journal of Research*, 15394492251328310. https://doi.org/10.1177/15394492251328310

Whiteside, D. M., Basso, M. R., Naini, S. M., Porter, J., Holker, E., Waldron, E. J., Melnik, T. E., Niskanen, N., & Taylor, S. E. (2022). Outcomes in post-acute sequelae of COVID-19 (PASC) at 6 months post-infection Part 1: Cognitive functioning. *The Clinical Neuropsychologist, 36*(4), 806–828. https://doi.org/10.1080/13854046.2022.2030412

Wieczorek, M., Reinecke, R., Borrat-Besson, C., Meier, C., Haas, M., Ihle, A., Kliegel, M., & Maurer, J. (2024). Cognitive functioning and sustained internet use amid the COVID-19 pandemic: Longitudinal evidence from older adults in Switzerland. *Scientific Reports, 14*(1), 18815. https://doi.org/10.1038/s41598-024-69631-w

Winter, D., & Braw, Y. (2022). COVID-19: Impact of diagnosis threat and suggestibility on subjective cognitive complaints. *International Journal of Clinical and Health Psychology, 22*(1), 100253. https://doi.org/10.1016/j.ijchp.2021.100253

Wu, X., Li, J., & Li, Y. (2022). The impact of uncertainty induced by the COVID-19 pandemic on intertemporal choice. *Journal of Experimental Social Psychology, 103*, 104397. https://doi.org/10.1016/j.jesp.2022.104397

Xiao, H., Zhang, Y., Kong, D., Li, S., & Yang, N. (2020). Social capital and sleep quality in individuals who self-isolated for 14 days during the coronavirus disease 2019 (COVID-19) outbreak in January 2020 in China. *Medical Science Monitor, 26*, e923921. https://doi.org/10.12659/msm.923921

Xiao, H., Zhang, Z., & Zhang, L. (2022). A diary study of impulsive buying during the COVID-19 pandemic. *Current Psychology, 41*(8), 5745–5757. https://doi.org/10.1007/s12144-020-01220-2

Xie, Y., Xu, E., & Al-Aly, Z. (2022). Risks of mental health outcomes in people with covid-19: Cohort study. *BMJ, 376*, e068993. https://doi.org/10.1136/bmj-2021-068993

Xiong, J., Lipsitz, O., Nasri, F., Lui, L. M. W., Gill, H., Phan, L., Chen-Li, D., Iacobucci, M., Ho, R., Majeed, A., & McIntyre, R. S. (2020). Impact of COVID-19 pandemic on mental health in the general population: A systematic review. *Journal of Affective Disorders, 277*, 55–64. https://doi.org/10.1016/j.jad.2020.08.001

Yaribeygi, H., Panahi, Y., Sahraei, H., Johnston, T. P., & Sahebkar, A. (2017). The impact of stress on body function: A review. *EXCLI Journal, 16*, 1057–1072. https://doi.org/10.17179/excli2017-480

Yelin, D., Margalit, I., Nehme, M., Bordas-Martínez, J., Pistelli, F., Yahav, D., Guessous, I., Durà-Miralles, X., Carrozzi, L., Shapira-Lichter, I., Vetter, P., Peleato-Catalan, D., Tiseo, G., Wirtheim, E., Kaiser, L., Gudiol, C., Falcone, M., & Leibovici, L. (2022). Patterns of long COVID symptoms: A multi-center cross sectional study. *Journal of Clinical Medicine, 11*(4), 898. https://doi.org/10.3390/jcm11040898

Ying, W., & Cheng, C. (2021). Public emotional and coping responses to the COVID-19 infodemic: A review and recommendations. *Frontiers in Psychiatry, 12*, 755938. https://doi.org/10.3389/fpsyt.2021.755938

Zager Kocjan, G., Kavčič, T., & Avsec, A. (2021). Resilience matters: Explaining the association between personality and psychological functioning during the COVID-19 pandemic. *International Journal of Clinical and Health Psychology, 21*(1), 100198. https://doi.org/10.1016/j.ijchp.2020.08.002

Zeraatkar, D., Ling, M., Kirsh, S., Jassal, T., Shahab, M., Movahed, H., Talukdar, J. R., Walch, A., Chakraborty, S., Turner, T., Turkstra, L., McIntyre, R. S., Izcovich, A., Mbuagbaw, L., Agoritsas, T., Flottorp, S. A., Garner, P., Pitre, T., Couban, R. J., & Busse, J. W. (2024). Interventions for the management of long covid (post-covid condition): Living systematic review. *BMJ, 387*, e081318. https://doi.org/10.1136/bmj-2024-081318

Zhang, H., Yang, Y., Hao, X., Qin, Y., & Li, K. (2023). Effects of digital sleep interventions on sleep and psychological health during the COVID-19 pandemic: A systematic review and meta-analysis. *Sleep Medicine, 110*, 190–200. https://doi.org/10.1016/j.sleep.2023.07.036

Zhaoyang, R., Scott, S. B., Smyth, J. M., Kang, J. E., & Sliwinski, M. J. (2020). Emotional responses to stressors in everyday life predict long-term trajectories of depressive symptoms. *Annals of Behavioral Medicine, 54*(6), 402–412. https://doi.org/10.1093/abm/kaz057

Zhou, H., Lu, S., Chen, J., Wei, N., Wang, D., Lyu, H., Shi, C., & Hu, S. (2020). The landscape of cognitive function in recovered COVID-19 patients. *Journal of Psychiatric Research, 129*, 98–102. https://doi.org/10.1016/j.jpsychires.2020.06.022

Psychological Mechanisms of Collective Mental Time Travel During the COVID-19 Pandemic

Ceren Dönerkayalı, Lynn Ann Watson, and Sezin Öner

Abstract This chapter explores how collective mental time travel (MTT)—the shared ability to mentally reconstruct the past and imagine future scenarios at the group level—was shaped during the COVID-19 pandemic. Drawing on psychological theories of memory and identity, the chapter delineates how shared representations of collective past and future emerged during this global crisis. Through data from a large international study involving over 4000 participants across 15 countries, the chapter demonstrates the convergence and stability of pandemic-related themes—such as lockdowns, infections, and economic consequences—across temporal (past, future) and spatial (national, global) dimensions. Notably, pandemic-related collective future thoughts exhibited greater positivity compared to past events, especially at the national level. Individual differences, such as affective traits and personality factors, modulated the emotional tone and frequency of spontaneous collective memories and future thoughts. Moreover, the chapter highlights the role of media narratives and political discourse in shaping the accessibility and emotional resonance of collective event representations. Counterfactual thinking and spatial optimism further influenced how groups imagined alternative pandemic trajectories and perceived threat severity. Importantly, collective identity—whether political, religious, or national—played a central role in shaping the content, phenomenology, and emotional evaluation of collective representations. This body of evidence not only enriches our understanding of collective memory and prospection under crisis but also raises critical questions about the long-term impact of identity-based divergences, media framing, and social narratives on collective event construction. Future research should investigate how concurrent events and social movements (e.g., the Black Lives Matter protests) interact with collective representations and how memory for such global crises evolves over time. Understanding

C. Dönerkayalı · S. Öner (✉)
Department of Psychology, Kadir Has University, Istanbul, Turkey
e-mail: ceren.donerkayali@stu.khas.edu.tr; sezin.oner@khas.edu.tr

L. A. Watson
Department of Psychology and Behavioural Sciences, Aarhus University, Aarhus, Denmark
e-mail: lynn@psy.au.dk

M. Migueles Seco, A. Aizpurua Sanz (eds.), *Autobiographical Memory and the Covid-19 Pandemic*, https://doi.org/10.1007/978-3-032-03895-1_3

collective MTT is essential for improving public communication strategies and fostering social cohesion during future global challenges.

Keywords Collective memory · Mental time travel · COVID-19 pandemic · Future thinking · Social identity

1 Introduction

Halbwachs (1925, 1992), the sociologist who initially conceptualized *collective memory*, defined it as memories of events shared by a group that are rooted in the group's socially structured history. While some definitions emphasize memories shared by a community and downplay individual memories of events experienced by community members (e.g., Assmann & Czaplicka, 1995; Olick, 1999), Halbwachs (1925) argued that individual memories are grounded in collective memories since, in broad terms, our life stories are at least a piece of a socially structured history. Individual or autobiographical memories are unique to those who experience them. This uniqueness does not arise from a disconnection with the community, but rather from each individual's specific position within that community (Halbwachs, 1925; also see Heux et al., 2023). Based on this, individual memories play a central and dynamic role in the formation and development of collective memory. This conceptualization is reflected in more recent definitions that describe collective memories as individual memories that are shared by a community, which in turn influences the identity of that community (Hirst & Manier, 2008).

The COVID-19 pandemic was a global crisis that not only disrupted daily life but also transformed how societies remember the past and anticipate the future. In line with current definitions, collective mental time travel refers to remembering memories or imagining future events that belong to a social group that includes the individual as a member (Michaelian & Sutton, 2019), and in this chapter, we examine how this process unfolded in the context of the COVID-19 pandemic. Drawing from psychological theories and empirical research, we explore the conceptual foundations of collective memory and future thinking, and present findings from a large-scale international study conducted across 15 countries, which traced the thematic and phenomenological characteristics of collective representations during different phases of the pandemic. We further consider the roles of media and political narratives in shaping collective recollection and prospection and, investigate cognitive biases such as counterfactual thinking and spatial optimism. Finally, we address how individual differences, including personality, affective traits, and social identity, influence the content and tone of collective mental time travel. Through this multifaceted analysis, we aim to deepen our understanding of how collective memory and future thought evolve under conditions of widespread uncertainty and shared societal threat.

2 Revisiting the Past, Projecting the Future: Mental Time Travel in Individuals and Groups

Our ability to represent ourselves at different time points in our subjective timeline is called *mental time travel*, and includes both past and future perspectives (Tulving, 1985). Studies have revealed that the past and future are interconnected for at least two reasons. First, past experiences are flexibly recombined to lay the foundation for the construction of future event simulations (Schacter & Addis, 2007). Second, similar brain regions are activated when individuals retrieve memories and engage in future simulations (e.g., Addis et al., 2007, 2009). Although Tulving's initial conceptualization of mental time travel refers only to individual memories and future events, Szpunar and Szpunar (2016) proposed that this connection may also extend to the collective domain. In line with this, a number of studies have identified similarities in perceived agency, phenomenology, episodic specificity, emotional valence (Topçu & Hirst, 2020), and content (Öner & Gülgöz, 2020) across collective past and futures. When differences between past and future thinking are observed in collective representations, these often-parallel differences are observed in individual memory (Topçu & Hirst, 2022). Researchers have thus suggested that exploring the concept of mental time travel for collective events may further our understanding of the interplay between cognitive and emotional processes that shape memory at the individual level, and collective event representations, narratives and identities more broadly (De Saint-Laurent, 2018; Merck et al., 2016; Michaelian & Sutton, 2019; for a review see Topçu & Hirst, 2022). In line with current definitions, *collective mental time travel* here refers to remembering memories or imagining future events that belong to a social group that includes the individual as a member (Michaelian & Sutton, 2019).

The COVID-19 pandemic was a sudden, unpredictable, and global outbreak that initially emerged in Wuhan, China, in December 2019. According to Johns Hopkins Coronavirus Resource Center, over 676 million cases have been reported, and more than 6 million lives lost (CoronaVirus Resource Center, 2023). Even if individuals did not experience coronavirus directly or lose loved ones because of it, the pandemic had psychological and physical impact across the world. Calls to action (i.e., World Health Organization, 11 March 2020) were made for country leaders to implement severe preventative actions. Some of these involved domestic and international travel restrictions, school closures, remote working, lockdowns, social distancing, and mandatory mask wearing (Hiscott et al., 2020). The majority of collective events affect relatively small groups of individuals; such as the coup attempt against the government on the 15th of July, 2016 in Turkey, or the attack on the Capitol in the US on the 6th of January, 2021, one of the unique characteristics of the COVID-19 pandemic is that it had worldwide impact. As such, the pandemic was represented heavily in individual, national, and global mental time travel during its occurrence and became a point of collective reference (Fridman & Gensburger, 2023; Öner et al., 2023).

Examining collective mental time travel in the context of the COVID-19 pandemic provides us with a unique opportunity to explore the connection between collective memories and collective future events at global as well as national levels. This is important because comparisons of collective mental time travel between collective identities have received relatively less attention than comparisons between individual and collective mental time travel. In this chapter, we aim to explore collective mental time travel during COVID-19. We will begin by explaining the conceptual foundations of collective memory and collective future thinking, and then present research investigating the themes and phenomenological characteristics of collective MTT during the COVID-19 pandemic. The chapter will focus primarily on findings from a large international study conducted across 15 countries, 4000 participants and two time points (see https://osf.io/m46nq/ and https://osf.io/5dazv/). The final section will consider individual differences in collective mental time travel.

3 Conceptual Foundations: Collective Memory and Future Thinking

Tulving (1985) defined the ability of people to project themselves at different time points in their subjective timeline extending from the past to the future as *mental time travel*. To engage in mental time travel, there are three essential requirements. The first of these is that there must be a traveler, a *self* that engages in this mental travel. The second is the presence of *autonoetic consciousness*, an awareness of the memories as one's own experience, in other words, a subjective sense of reliving the memory. Last, to move backward or forward in time, there must be a *subjective timeline* that extends from the past to the future within the individual mind. The physical time when an event happened in the past and the time point where this event is placed on the subjective timeline may not align, since, as the name suggests, this is a unique timeline that belongs to the individual (Tulving, 2002).

To some degree, conceptualizations of individual and collective mental time travel are similar since in both cases, what is done is merely to remember the past and simulate the future. However, several studies show that while some features overlap, other underlying features differ (see Szpunar & Szpunar, 2016). During autobiographical event processing, individuals with hippocampal damage are unable to (re)construct past and future events (Hassabis et al., 2007; Tulving, 1985), yet their ability to generate collective past and future events remains intact (Klein et al., 2002; Szpunar et al., 2014). Additionally, when the three conditions required for individual mental time travel are examined, it can be seen that these requirements are not met in collective mental time travel. First, the concept of *self*-differs, while the traveler in episodic thinking is the self that engages in mental time travel, in collective memory the self-component activated refers to the identity the individual shares as part of a group, religion, nation, or culture. Second, there is no need for *autonoetic consciousness* or a *subjective timeline* within the individual's own lifetime when (re)constructing collective memories or future events (Michaelian & Sutton, 2019). In contrast, collective mental time travel can then refer to events the individual has personally experienced (i.e., the closure of schools and workplaces following a governmentally enforced lockdown during the pandemic) or events that they have not experienced (i.e., a young person who was raised in New York and identifies as a "New Yorker" recalling a narrative about the attacks on the World Trade Center on September 11th, 2001), the defining criteria being that these events occurred for a social group of which the individual is a member.

Further evidence that touches upon the differences between individual and collective mental time travel relates to *temporal orientation preference,* that is, a general preference for attending to the past, present, and future (Nuttin, 1985; Zimbardo & Boyd, 1999). While individuals can focus on all temporal orientations to a certain extent (e.g., Shipp et al., 2009), some studies demonstrate that individuals are likely to focus more heavily on one temporal orientation (e.g., Holman & Silver, 1998), such as dominantly preferring a past-oriented focus over the present and future. Temporal orientation preference in collective mental time travel is largely disregarded. Although personal temporal focus is emphasized in individual temporal orientations, the essential detail in collective temporal orientations is that they include a temporal focus shared within a group (Peetz & Woohl, 2019). In fact, each social group might have a distinct pattern in its collective temporal orientations of past, present, and future. For instance, in a national context, the US may be more present-oriented since it emphasizes immediate concerns and gratifications. In contrast, Germany might be more future-oriented in prioritizing the development of the next generation (Geiger et al., 2021). For the global perspective, each international organization or community follows a unique temporal focus for the global crisis, such as pandemics, climate change, or global economic crises. Hence, national and global temporal orientations are influenced by shared representations of the collective memories and collective future events, and this, in turn, can influence the decisions and behaviors of nations and the world, manifesting itself in policy-making and international collaboration. For example, shared representation of Germany regarding the value of education of future generations might adopt a future temporal orientation and, as a result, may make greater investments in education.

The concept of collective memory has been investigated extensively across a number of different sub-disciplines beyond psychology, such as sociology, anthropology, and history and therefore, it is challenging to find a common definition that satisfies all disciplines (Heux et al., 2023). However, in psychology, researchers generally agree upon two components that should be included in the definition of collective mental time travel. First, as outlined above, collective mental time travel should include a *shared representation* of the past or future. Second, there must be an interrelationship between the shared representation and the *group's identity* (Coman et al., 2009; Hirst & Manier, 2008; for a review, see Stone & Jay, 2019). While shared representations of the past refer to socially structured symbols, cultural artefacts, narratives, or practices that help group members commemorate the collective event in memory (Hirst & Manier, 2008, for a review, see Coman et al., 2009; Stone & Jay, 2019), collective future shared representations refer to socially structured predictions, expectations, projections, and narratives about the future (Szpunar & Szpunar, 2016). Importantly, a dynamic relation exists between the collective past and future: shared representations of the collective past establish a foundation for imagining the group's collective future (Szpunar & Szpunar, 2016) and, shared representations of the future can also inform the collective past to maintain the continuity of the collective identity (Sani et al., 2007; Szpunar & Szpunar, 2016).

According to Hirst and Manier (2008), the answer to the question of why only some memories become collective memories is based on mechanisms of *transmission*, *convergence*, and *stability*. Stone and Jay (2019) also suggest the additional mechanism of *social identity*. These four mechanisms can also be adapted to inform our understanding of collective future thinking. Transmission refers to the act of discussing or sharing a public event or future expectations with other members of a social group. The recollections of a public event or expectations for the future may not be the same for each group member, yet ongoing discussion of these events is likely to maintain accessibility for collective representations over time (Hirst et al., 2018). In fact, as group members continue to discuss events over time, their collective event representations become more homogenous, a phenomenon labelled *mnemonic convergence* (Hirst & Manier, 2008; Michaelian & Sutton, 2019; for reviews, see Hirst & Coman, 2018; Stone & Jay, 2019). Study findings emerging from the social contagion paradigm (e.g., Meade & Roediger, 2002) support mnemonic convergence through the transmission of information through social interaction, and socially shared retrieval-induced forgetting (SSRIF, e.g., Cuc et al., 2007) observed both in the speaker and the listener (Hirst & Coman, 2018; Stone & Jay, 2019). In addition, to ensure that public event representations will remain stable through future generations, they may also be included in books, social narratives, or monuments, and museums and other cultural artefacts (Hirst & Manier, 2008; Stone & Jay, 2019; Szpunar & Szpunar, 2016). Last, the social identity of the group has to remain connected to the public event or future expectations (see Coman et al., 2009; Stone & Jay, 2019; Szpunar & Szpunar, 2016). In line with this Abel et al. (2019) demonstrated that while the collective memories of the events of World War II were relatively similar across 11 countries (i.e., D-day, the attack on Pearl Harbour, the holocaust), differences were found for events that had a significant impact on a specific country (i.e., the Battle of Britain in the UK, the Battle of Stalingrad in Russia, and De-Gaulles appeal in France).

4 COVID-19 as a Catalyst for Collective Mental Time Travel

Here, we focus on common themes and the phenomenological characteristics (e.g., vividness, emotional intensity, centrality, etc.) of collective pandemic events, and the potential role of media and political narratives.

4.1 The Pandemic as a Collective Event

Research shows that emotional events are more likely to be shared among community members than less emotional events (e.g., Harber & Cohen, 2005; Luminet et al., 2000) and emotional intensity predicts to what extent individuals discuss such public events with others (Rimé et al., 1998) making these events more likely to be transmitted through a social group and converge into a collective memory maintained by a social group (Hirst & Manier, 2008). Since the pandemic was a highly emotional global event, the transmission of information and events related to it was inevitable, as evidenced by the continual and massive presence of pandemic-related events on mass media and social media during the pandemic itself. In their edited collection, Fridman and Gensburger (2023) defined the "COVID-19 memory boom" as accumulated knowledge that began with the onset of the pandemic. They suggested that it aids researchers in referring to the COVID-19 pandemic as a collective memory.

Öner et al. (2023) aimed to examine convergence in common collective event themes during the early phase of the COVID-19 pandemic, between April and June 2020 across spatial (national and global) and temporal orientations (past and future). They asked more than 4000 participants who were citizens from 15 different countries to report on significant memories and future expectations about events that had occurred or were expected to occur in their country or around the world. For each of the four different event categories (i.e., past-national, future-national, past-global, future-global), participants reported three events. In general, the results demonstrated that nearly 85% of events across all event categories referred to the pandemic. When looking at the findings in more detail, they found that themes were shared both within and between countries. In collective memories, events related to lockdown (>10%), infections (>9%), politics (>6%), and health systems (>4%) formed the majority of themes for national as well as global orientations. For global memories, themes of "death" (5.7%), "cultural events" (4.9%), and "transportation restrictions" (4.2%) were also frequent whereas the themes of, "mass closures" (8.9%), and "hygiene and social behavior" (6.1%) were more common in national memories. For future events, themes were shared to an even greater extent across national and global orientations. The most common theme in national and global futures being "impact on the economy" (>22%), and the possibility of a "second wave" (>4%). For global future events, themes of health science (5.0%), politics (5.0%), and travel (4.8%) were common whereas for national events, the possibility of future "lockdowns" (12.4%) was evident.

Besides this convergence between the themes within and between nations during the early phase of COVID-19, Hirst and Manier (2008) also highlighted that *stability* is central for whether the COVID-19 pandemic will become a part of our history, as this will ensure the endurance of these shared representations through of the passage of time. To examine event stability, Watson et al. (2025) investigated themes in the past and future from both national and global perspectives, at different time points; one during pandemic onset and the other at a later phase. The study is a follow-up of Öner et al. (2023) and included over 4000 participants from 10 different countries. In addition to the data collection outlined above in 2020 (early phase), the authors completed an additional round of data collection between June and October 2022 (late). They found that the percentage of events reported that included themes related to the COVID-19 pandemic reduced from nearly 79% in the early phase to 33% in the late phase. Although this 46% reduction was observed, when the pandemic was mentioned, the content of the common themes remained stable. Themes of "infection" (early = 14.7%, late = 10.1%) and "lockdowns" (early = 15.3%, late = 5.9%) remained the most frequent pandemic themes across global and national memories. The theme of infection was most frequent at global level whereas lockdown was most frequent at the national level. A small increase in memories related to the end of the pandemic was also observed in "developments in health science (early = 0.6%, late = 4.1%). For future events, a 61% reduction in pandemic-related events was observed across the two time points. The most frequent theme for future events at both time points was the "end of the pandemic" (early = 10.7%, late = 6.2%), whereas a large reduction in frequency was observed in the pandemic themes related to the "impact on economy" (early = 23.8%, late = 2.1%).

During the late phase of the pandemic, when vaccines had been made available and the acute threat of the pandemic reduced, pandemic-related themes were replaced by non-pandemic themes. At the global level, the themes of "war," "environment," and "politics" became common, reported in more than 10% of events across 8 countries. At the national level the themes of politics and economy were most prevalent, also reported in more than 10% of events across 8 countries. Together, the results of these two studies are consistent with others showing a representational *convergence* across nations for global events (e.g., Abel et al., 2019, 2022). Additionally, Watson et al. (2025) is one of the first to examine the content of collective event representations for a worldwide health crisis across multiple time points. The findings indicate convergence and stability in the types of pandemic-related themes that occur over time, with themes relating to pandemic transmission common in memory and the end of the pandemic in future thought. However, the frequency of these themes reduced substantially over time as the threat of the pandemic lessened. In line with previous studies, collective events with greater socio-political impact began to dominate, and themes of war, politics, and economy emerged (e.g., Liu et al., 2005; Pennebaker et al., 2006; Schuman & Rodgers, 2004).

More specifically, Watson et al. (2025) demonstrated a clear shift in the thematic content of non-pandemic collective events between the early and late phases of the pandemic. During the early phase (spring 2020), the collective representation of events was overwhelmingly dominated by pandemic-related themes, with non-pandemic content comprising only a small proportion of reported events. During the late phase of the pandemic, non-pandemic themes such as politics and war became prominent in memory, with politics comprising 23.5% of global and 17.8% of national past events, and war accounting for 21.1% of global past events. These patterns were relatively consistent across countries, although national-level variations were observed; for instance, war was the most frequently reported national past event in Russia (18.4%), the majority of memories referring to the Russian invasion of Ukraine in spring 2022. Likewise, non-pandemic themes further diversified for the future events during the late phase, with war, environment, politics, and economy emerging as dominant concerns. Globally, war was the most frequently cited future event (19.6%), while politics was the most frequently cited among national future projections (31.8%). Country-level differences also emerged, such as the end of the pandemic remaining prominent in China (25.1%). Overall, these findings support the dynamic nature of collective memory and future thinking, showing that while the frequency of pandemic-related representations reduced across the pandemic, stability was evident across the most common themes in past and future thinking. This indicates that individuals converge over time on the themes that are central to our understanding of the pandemic as a collective event. Importantly, the content of the pandemic representations does not necessarily change, but rather collective representations evolve in response to shifting sociopolitical and global contexts when other events of potential collective significance occur.

4.2 The Phenomenology of Collective Pandemic Events

Phenomenological characteristics (e.g., emotional valence, vividness) of an event can help us to understand how the collective past and future is imagined. Comparing the emotional valence of personal past and future with the nation's past and future gives a clue about the distinctions between individual and collective mental time travel. Studies show that while there is a positivity bias in remembering the personal past and even more so imagining the personal future, a negativity bias can be observed in collective mental time travel. For example, Shrikanth et al. (2018) adapted the future fluency task to measure the accessibility of future events across different time periods (e.g., next week, next year, and between 5 and 10 years). For personal and collective future events, participants were instructed to list things that they were worried and excited about their future and their country's future within a 1-min interval. Findings showed that participants generated more positive events in their own future and more negative events in their nation's future. Several other studies have also found a similar pattern that is, anticipating more negative than positive events in collective future thinking (e.g., Hacıbektaşoğlu et al., 2023; Öner & Gülgöz, 2020; Shrikanth & Szpunar, 2021; Yamashiro & Roediger, 2019). Still, it should be noted this valence effect is not always consistent. Deng et al. (2023) did not replicate this pattern in participants in China by using a similar future fluency task, Topçu and Hirst (2020) found neutral valence in US. Mert et al. (2023) revealed that individuals from China were more positive about the long-term future of their country than participants from the US and Turkey and demonstrated that individuals experience of national well-being and identification with the values of their country may play a role in shaping their emotional expectations.

Phenomenological differences can also be observed between events generated from different collective perspectives, such as from national and global perspectives. Öner et al. (2023), investigated the emotional valence and vividness (past temporal orientation only) of national and global events. As in previous studies, they found collective future events were rated as less negative than past events, but specifically, this difference was more pronounced for national events. National past events were also rated as more vivid than global past events. Since this study was conducted during the early stage of the pandemic, the positivity bias in national future events and strong imagery for national past events may have emerged due to several dynamics. Öner et al. (2023) found that the differences between the valence and vividness of collective events across national and global perspectives were maintained when controlling for the severity of the pandemic and stringency of governmental restrictions across countries, suggesting that objective measures of pandemic impact did not account for these differences. However, the findings of Mert et al. (2023) indicate that individual difference factors such as national well-being and country identification may also play a role. Sociocultural factors such as the political narratives promoted during the pandemic and the portrayal of the pandemic in the media may additionally influence the phenomenology of collective representations, these factors will be discussed further in Sect. 4.3.

To answer the question of whether these biases are stable across different phases of pandemic, Watson et al. (2025) also examined emotional valence and vividness (past temporal orientation only) in pandemic and non-pandemic events across temporal (past, future), spatial (national, global), and time (early, late) dimensions. The results demonstrated that collective memories, (both pandemic, and non-pandemic related, across national and global orientations, during both the early and late phases of the pandemic) were mildly negative. In contrast, future events were generally mildly positive. Indicating, as in previous studies, that the collective future is more positive than the collective past (Hacıbektaşoğlu et al., 2023; Topçu & Hirst, 2020). No differences were found between pandemic and non-pandemic events during the early phase of the pandemic. However, during the late phase, as the threat of the pandemic was less acute, pandemic-related future events were more positive than non-pandemic future-related events. Importantly, in this study, pandemic severity within countries did account for differences in emotional valence between pandemic and non-pandemic-related events. However, differences observed in the emotional valence of global and national events were independent of pandemic severity. National events were rated as less negative than global events in the later phase of the pandemic and, consistent with Öner et al. (2023) national events were more positive and global events in future thinking only. Overall, the findings demonstrate stability in the positivity biases observed for future relative to past temporal orientations and national relative to global spatial orientations in future thinking only. In general, pandemic-related events were similar in emotional valence to non-pandemic-related events. Differences were only observed in future thinking during the late phase of the pandemic, and these were accounted for by differences in the impact of the pandemic severity at the country level. Other studies have also demonstrated that the emotional valence of pandemic-related memories may become more positive (e.g., Hjuler et al., 2025) and future trajectories about the pandemic more optimistic as individuals become less preoccupied over time (e.g., Topçu & Hirst, 2024).

In terms of vividness, national events were more vivid than global events across both time points and pandemic-related memories were rated as more vivid than non-pandemic memories during the early phase only. These findings were not related to pandemic severity. Taken together, the research demonstrates stability in the finding that national events are more phenomenologically vivid than other types of collective representations and yet that the vividness of collective representations can also be influenced by immediate social factors that also change over time.

4.3 Potential Role of Political Narratives

From the first COVID-19 case, there was a great deal of information (also misinformation, and disinformation, Hiscott et al., 2020; Gottlieb & Dyer, 2020) and updates provided by governments (e.g., Ministry of Health), mass media (e.g., television), social media, and global organizations (e.g., World Health Organization). For instance, in Turkey, it became routine during the pandemic for the Minister of Health to appear on television every evening to announce the daily number of COVID-19 cases and COVID-related deaths. Information about national and global events related to the pandemic that were covered by the media stream have a reciprocal relationship with political narratives regarding the pandemic. Political leaders, governments, and global organizations can create political narratives to shape shared representations and public opinion regarding the event or crisis and to justify and generate support for the policies they enact. For example, political narratives about COVID-19, such as perception of the severity of the pandemic, and how stringent the governments' attitudes, restrictions, or reactions were, in turn affected the COVID-19-related media stream (e.g., Soyaltin-Colella & Sert, 2024). Media reinforcement and socially selectively induced forgetting mechanisms potentially lead to mnemonic convergence or *collective forgetting* through conversational remembering, and may further maintain political narratives expressed through mass or social media (for reviews, see Hirst & Coman, 2018; Stone & Jay, 2019).

Before looking at how media and political narratives influence the memories and future expectations regarding the pandemic, we will focus on the potential role of media in collective memories and collective future events to better elucidate the role of these mechanisms. Schuman and Rodgers (2004) asked American participants to report significant events that took place in the history of the world, across three time points: 1985, 2000, and 2001 (after the 9/11 attack). Their findings indicated that although World War II was consistently reported at all time points, shared representations were replaced when highly emotional, unpredictable, and impactful events such as the 9/11 attack. Researchers found that at the 2001 time point, participants reported the 9/11 attack more frequently than World War II, indicating that some events became less accessible as more recent, higher profile events became a priority. Schuman and Rodgers (2004) further stated that accessibility of collective events might depend on whether an individual directly experiences this collective event and the extent to which these events received extensive media coverage. Similar dynamics between individual difference variables, mass media, narratives, and local or global institutions, are also hypothesized to shape collective future thinking (Szpunar & Szpunar, 2016).

In their study, Öner et al. (2023) also examined whether themes differ across nations based on severity and stringency measures, which referred to the number of total cases per million and the governmental stringency index, respectively, for each country. For the past global events, when pandemic severity was low and governmental stringency high, participants reported higher numbers of event themes related to "infection" and "politics." For national past events, "lockdown-related events" were most frequent, regardless of the level of severity and stringency. In addition, the themes of "social solidarity" and "health" were more common in low-severity or high-stringency countries. Taken together, these findings indicate that pandemic severity and stringency had a direct impact on the themes evident in collective memories of the pandemic. Based on the literature outlined above, the content of these themes are likely to have been influenced by media and political coverage within each country. For example, in a country where pandemic severity was low, the media stream was potentially dominated by the news about infections in other countries to maintain the social narrative that pandemic is a severe worldwide threat, and that individuals within that country should follow enforced governmental restrictions. The theme of lockdown at the national level was common regardless of pandemic severity and stringency, which may indicate the centrality of this intervention to the management of the pandemic and the importance of media outlets in transmitting information about the implementation and impact of national strategies. In high-stringency countries, themes of health and social solidarity were also common and generally positive. This may reflect additional media coverage of social narratives promoting positive messages of hope to maintain social distancing behaviors under conditions where policies were highly restrictive.

When looking at future events, the "economy" was by far the most common theme. However, event themes related to the "economy" were more salient in low-severity and high-stringency countries as well as the end of the pandemic at the national level. For high-severity and low-stringency countries, the potential for future lockdowns was more frequent at both the global and national levels as well as the potential for a second wave at the national level. Öner et al. (2023) highlighted that since the long-term consequences of the pandemic were ambiguous during 2020, future expectations regarding the global and national stances can be based on conversational acts of remembering with groups (e.g., friends, family, etc.) or media coverage (also see Anderson, 2012). For countries where governmental restrictions were stringent, social narratives reflecting the impact of the pandemic on the economy are important to allow discussion of how this impact will be addressed in the future, the higher representation of themes related to the end of the pandemic theme may also reflect the promotion of the belief that such restrictions will be effective. In contrast, where pandemic severity was high and governmental stringency low, future representations may reflect social narratives that future lockdowns may yet be implemented and the extent to which a second wave of the pandemic may occur.

5 Collective Thinking and Public Health Messaging

A large body of research is dedicated to understanding how cognitive biases influence decision making and behavioral change (e.g., Tversky & Kahneman, 1974). Collective behavioral change was central to the majority of the government strategies implemented worldwide to manage the pandemic such as social distancing, travel restrictions, mask wearing, and hand washing. This section of the chapter examines features of collective mental time travel that may influence the effectiveness of public health messaging such as collective counterfactual thinking and spatial optimism.

5.1 Scenario Thinking and Counterfactuals

Counterfactual thinking can be defined as thinking about alternative scenarios for events that could have happened but did not (Roese & Olson, 1997). Individuals can generate numerous counterfactuals for their memories, engaging in this process is not effortful; instead, it only requires thinking about the question of "What if?" (Byrne, 2016). There are two common types; upward and downward counterfactuals. While *upward counterfactual thinking* involves thinking about better alternatives, *downward counterfactual thinking* involves thinking about worse outcomes (Roese, 1994). An upward counterfactual during the pandemic may be imagining that you did not get infected by COVID-19, even if you did. Conversely, a downward counterfactual based on the same scenario may be imagining that you spread the infection to your elderly parents, even if that was not the case. When applying counterfactuals at the collective level, an upward alternative might be imagining how much better it could be if the lockdown or travel restrictions had been applied earlier in your country, whereas a downward alternative might be imagining a scenario in which the number of cases and deaths was much higher than actually reported in your country. In addition to influencing the event's emotional experience by creating alternative worlds, counterfactuals can be used as a source for future trajectories.

Research investigating counterfactual thinking in the context of the pandemic has demonstrated that actively asking individuals to engage in counterfactual thinking about the outcome of the pandemic prior to engaging with pandemic-related messaging from media news outlets or governmental sources can lead to less engagement with fake news headlines in those with high levels of conspiracy mentality (Bertolotti & Catellani, 2023) and reductions in the polarization of attitudes towards COVID-19 prevention behaviors in individuals with differing political orientations (García Ferrés & DePalma, 2023).

To examine the effect of counterfactuals on public health messaging more specifically, Seifert and Somasundaram (2025) provided individuals with visual and written information about the daily confirmed number of COVID-related deaths since the onset of the pandemic. Individuals were then asked to engage in counterfactual thinking pertaining to an alternative outcome for the evolution of the death toll either before (treatment group) or after (control group) completing a forecasting task estimating the death toll for the following day in the US.

The results demonstrated that the majority of individuals (>95%) generally employed upward counterfactual reasoning, that is imagining more positive outcomes. In line with this, individuals also generally underestimated future pandemic-related outcomes across both groups. However, this underestimation was larger in the treatment group who engaged in the counterfactual thinking prior to the forecasting task. Additionally, researchers found that underestimations were greater in individuals with longer counterfactual descriptions and in those who had themselves been infected or, had connections to someone who had been infected with the COVID virus. These findings indicate that individuals generally underestimate the impact of health-related crises in their own country and that this bias may be anchored in general optimism biases in future thinking, including the ability to generate upward counterfactuals about collective events. These optimism biases are further exacerbated in individuals who engage more elaboratively with counterfactual thinking or in those who have direct personal experience that the virus does not commonly result in death. The authors argued that these findings highlight the need to frame public health messaging in emotionally neutral terms to minimize the impact of optimism biases on public engagement with health-related messaging.

Studies of public health-related messaging during the initial phase of the pandemic in fact demonstrated that public messaging is actually often associated with threat appeals. These appeals include information about negative or unfavorable outcomes that may occur if the public do not act or intervene. While these threat appeals can be effective in eliciting an emotional response and psychological distress, they are less effective at increasing individuals' intentions to engage with behavioral interventions that can mitigate the effects of a health crisis (Richmond et al., 2022). In relation to this, Ahn et al. (2021) found that engagement with behavior interventions can be improved when if-statements, that may encourage downward counterfactuals, are combined with concrete plans about what do if the situation arises. They found that engagement with social distancing practices was higher for individuals who were asked to generate three concrete action plans about what to do in a specific situation X, if they were suddenly at risk of being infected with COVID-19. For example, "if I (X) had to go to the grocery store for my shopping I would (Y) put my mask on and ensure I kept my distance from other customers in the shop when waiting in the queue".

Taken together these findings indicate that engaging with counterfactual thinking about the pandemic can lead to emotional and cognitive biases that may reduce the effectiveness of public health messaging. To minimize the impact of these biases public messages should be clear, concise, and emotively neutral to reduce the likelihood of inducing the natural tendency to engage in upward counterfactual thinking. When messages include potentially negative outcomes that may lead to downward counterfactual reasoning, these should be supplemented with the opportunity for intention implementation where individuals are provided with or self-generate, specific plans to help them concretely manage risky situations if these are to arise.

5.2 Spatial Optimism

Spatial optimism refers to the tendency to underestimate the seriousness of an environmental threat when considering that it is happening in a national context compared to a global context (e.g., Coquet et al., 2019; Musson, 1974). There are several reasons why spatial optimism may be observed. First, individuals might underestimate threat severity at the national level simply because of group size. Individuals may be more optimistic about the outcome of a threat directed at a smaller relative to a larger group. In line with this, Fleury-Bahi (2008) examined whether risk perception based on 15 different environmental risks differs by group size. They found that the larger the size of the group, the greater the risks perceived. Specifically, the perceived risks of environmental threats were greater for humanity compared to a country, and the perceived risks were greater for a country compared to a town, which were then also greater than the perceived risks to oneself. Second, spatial optimism may reflect differences in knowledge or media exposure about the ongoing environmental threat. Gifford et al. (2009) found that spatial optimism is not universally evident for environmental threats, and that instead, estimates of environmental threat are influenced by awareness or local knowledge. Finally, spatial optimism may also reflect an ego-defensive strategy enhanced by collective narcissism. When collective groups feel threatened, they may exaggerate in-group achievements and attributes to maintain collective identity (Yamashiro et al., 2022). Spatial optimism is an important phenomenon as it has been found to influence behavior. When individuals believe that an environmental threat is not a serious problem, they are less likely to engage in behaviors to prevent or minimize the potential impact (see Baldassare & Katz, 1992).

In the context of the pandemic, spatial optimism may lead to individuals from one nation believing that the pandemic is less likely to spread to their country relative to other countries or that the impact of the pandemic may be less severe. As a result, they might not follow the restrictions enforced by their government to prevent the spread of the pandemic, such as mandatory wearing of masks, transportation restrictions, and social distancing. As an example, the first COVID-19 case in Turkey was detected in March 2020. Before the first case, there was a general public belief that COVID-19 might not spread to Turkey since Turkish people wash their hands very often.

Spatial optimism has primarily been investigated in relation to environmental threats (Tvinnereim et al., 2020) and to the authors knowledge, only one study has investigated the impact of spatial optimism on individuals' expectations about the infection rates and the duration of the COVID-19 pandemic. Öner et al. (2025) asked participants from 15 different countries about their estimations of (1) when the COVID-19 pandemic would be under control (after a few weeks to more than a year) and (2) the infection rate (0–100% of the population infected) for their country and the world. Spatial optimism was measured by comparing the country-level and global-level estimations. As hypothesized, participants indicated that infection rates would be lower, and the COVID-19 pandemic would come to an end sooner when estimating for their own country relative to the whole world. Importantly, the pandemic severity index moderated these effects. In line with previous studies of environmental threat, spatial optimism was greater in lower-severity countries, indicating that the bias was modified by local knowledge or media coverage about the ongoing threat at the national level This moderation effect suggests that people continuously update and adjust their pandemic-related beliefs based on new information from their national or collective environment (Tversky & Koehler, 1994). The important take-home message of the study is that spatial optimism can be observed during a collective event such as the COVID-19 pandemic. This highlights the importance of considering individual and collective expectations about the future shaping beliefs about collective threats. In terms of public health messaging, these findings demonstrate the importance of clear and precise communication strategies during public health crises and the need to disseminate accurate and regular messages at the national level to minimize optimism biases and maximize adherence to social interventions.

6 Individual Differences in Collective Mental Time Travel

In this section, we will review studies revealing how individual differences modulate collective mental time travel. Our focus will be on the role of personality (e.g., openness, neuroticism, etc.) and affective traits (e.g., trait anxiety, etc.) in shaping collective memory and future thought, and the impact of social identity and political ideology on collective memory representation and future expectations.

6.1 Age Differences in Collective Mental Time Travel

Age significantly shapes how individuals engage in collective mental time travel (MTT), particularly in terms of the emotional tone and cognitive framing of imagined future events. A robust finding across studies is the "positivity effect" observed in older adults, who tend to recall and imagine more positive than negative collective events, even during crises such as the COVID-19 pandemic. This age-related trend is attributed to motivational shifts described by socioemotional selectivity theory (Mather & Carstensen, 2005), which suggests that as individuals age, they prioritize emotionally meaningful and rewarding information. Aizpurua et al. (2021) found that older adults not only recalled more positive future statements but also frequently transformed negative content into more neutral or optimistic versions, reflecting a cognitive bias geared toward emotional regulation. Middle-aged adults showed similar, although less pronounced, patterns. Complementing this, Burnett et al. (2023) reported that older individuals emphasized themes of resilience and community strength in their collective future thinking, while younger adults were more likely to focus on uncertainty and institutional shortcomings. Similarly, Zaragoza Scherman et al. (2022) showed that although different age groups recalled similar pandemic-related themes, older adults framed these events with greater emotional positivity and interpretive coherence. These findings suggest that with age, collective MTT becomes increasingly guided by affective goals and regulatory strategies, helping individuals navigate adversity with greater emotional resilience. Future research should examine how these patterns interact with cognitive functions such as memory specificity and inhibition, and whether the positivity effect generalizes across other types of collective events beyond the pandemic.

6.2 Personality and Affective Traits

As with autobiographical memory, affective traits and personality characteristics play a crucial role in shaping how individuals encode, retain, and retrieve collective memories (Berntsen, 2021; Kensinger, 2009). Individuals with a tendency to experience heightened emotional states typically remember past events as emotionally charged, resulting in spontaneous memories that are frequent and emotionally intense (Shan et al., 2025). When individuals choose to remember past events, these memories are also more commonly negative, associated with psychological distress and reduced memory specificity (see Dalgleish & Hitchcock, 2023 for a review). While few studies have directly examined the impact of affective traits and personality characteristics on both autobiographical and collective mental time travel simultaneously, some studies of collective events such as studies investigating the terrorist attacks on America on the 11th of September, 2001, demonstrate that similar effects are also observed in collective memory. Memories of this event are found to be more negative in individuals reporting higher levels of psychological distress (e.g. Levine et al., 2005). Additionally, while negative emotions commonly fade over time, an effect known as the fading affect bias (Skowronski et al., 2014), this bias was not observed for memories of the terrorist attacks in individuals from the US with post-traumatic stress disorder, instead these individuals had a tendency to inflate the emotional aspects of their memory over time (Qin et al., 2003).

In the context of the pandemic, individuals with higher anxiety and depressive symptoms reported experiencing more frequent and emotionally negative spontaneous memories and prospective thoughts related to COVID-19 (Cole et al., 2023). These spontaneous thoughts were significantly influenced by the emotional salience of COVID-related experiences and attentional focus on pandemic-related information, highlighting the parallels between collective and autobiographical memory processes. This effect for pandemic-related thought mirrors how event representations in past and future thinking are more negative in clinical contexts (del Palacio-Gonzalez & Berntsen, 2020; Watson et al., 2012) and suggests that both personal and collective crises shape the formation of events representations and impact mental health.

The role of negative affectivity in memory has also been found to be further exacerbated by prolonged isolation and social distancing during COVID-19. Zhang et al. (2021) observed a U-shaped relationship between the duration of social distancing and memory errors, where extended isolation initially reduced memory errors but later increased them significantly alongside rising negative emotions such as anxiety, loneliness, and depression. This evidence suggests that prolonged emotional distress and loneliness during collective events modulate memory whereby negative emotional states impair memory accuracy, increase errors, and intensify negative emotional recall (Zhang et al., 2021).

Additionally, the severity of the physical impact of becoming infected with COVID has also been found to shape the emotional and cognitive features of memory. Research on individuals experiencing prolonged COVID-19 symptoms ("Long COVID") showed pronounced memory impairments in domains such as prospective memory, verbal fluency, and incidental learning, accompanied by elevated emotional distress, anxiety, and depressive symptoms (Espinar-Herranz et al., 2023). These findings demonstrate that the intensity of personal emotional experiences associated with COVID-19 impacts memory formation and emotional representation in a manner closely resembling autobiographical memory processes.

Taken together, research indicates negative emotional states influence the phenomenology of pandemic-related memories as observed for autobiographical memories in more clinical contexts (del Palacio-Gonzalez et al., 2017; Watson et al., 2012), demonstrating further parallels in how individual difference factors influence individual and collective mental time travel. While the pandemic was a global event, each individual experienced it differently, with massive individual differences in the psychological, social, economic, and physical impact of the pandemic over time (McBride et al., 2021). Research suggests that these factors influence the emotional and cognitive features of memory and, illuminate the need for additional research to investigate the influence of these factors on collective mental time travel as well.

6.3 Social Identity and Belonging

During the COVID-19 pandemic, social identities—particularly political, national, and religious—served as powerful frameworks for shaping how individuals and groups encoded and interpreted shared experiences (Hirst & Echterhoff, 2012; Liu et al., 2021; Bavel et al., 2020). These identities influenced not only what aspects of the pandemic were remembered, but also how those events were emotionally evaluated and morally framed (Coman et al., 2009; Klein et al., 2011). For example, lockdowns, protests, or public health messaging were not recalled uniformly; instead, they were encoded and consolidated in ways that reflected group-specific concerns, values, and worldviews. Social identity shaped what was considered memorable, how those events were emotionally appraised, and how often they were rehearsed in interpersonal communication and media exposure. This process is consistent with the mechanism of socially shared retrieval-induced forgetting (Cuc et al., 2007), whereby frequent rehearsal of identity-consistent narratives enhances memory for in-group-relevant information while diminishing recall for discrepant or out-group content.

Moreover, national identity influenced how different populations remembered their country's pandemic trajectory. In contexts where national narratives emphasized unity, resilience, or scientific achievement, individuals were more likely to recall the pandemic through a lens of collective efficacy and pride (Liu et al., 2021). In contrast, in nations with fractured political discourse or high institutional distrust, collective memory became more contested, with groups constructing divergent and sometimes oppositional recollections of the same events. These identity-driven memories are not trivial—they inform the psychological foundations of how societies will evaluate institutional performance, assign blame, and set future expectations.

Importantly, the influence of identity extended not only to retrospective memory but also to collective prospection. Research in collective future thinking shows that the same identity-based frameworks used to organize the past also guide simulations of possible futures (Szpunar et al., 2014). During the pandemic, groups anticipating continued decline, governmental failure, or moral decay often projected these fears into pessimistic collective futures—particularly when their identity group was politically marginalized or felt underrepresented in the management of the crisis. Conversely, groups aligned with institutional authority or dominant national narratives often envisioned recovery, progress, and unity. The emotional tone of these collective simulations—ranging from hope to despair—was thus shaped by identity-based positioning within the social and political landscape of the pandemic (Hacıbektaşoğlu et al., 2023).

Collective memory serves both an epistemic and a regulatory function: it provides coherent narratives that help individuals and groups make sense of the past, and it supports emotional regulation and future planning (Brown, 2021). Identity plays a crucial role in both functions by shaping what is included or excluded from the collective account and by influencing how emotionally charged those memories become. In the context of COVID-19, identity-based divergence in memory content and valence likely served to enhance group cohesion and worldview defense, but at the cost of shared understanding between groups—a dynamic that may have long-term implications for social trust and intergroup relations.

In summary, from the initial encoding of pandemic events to their emotional coloring, rehearsal, and integration into broader narratives, psychological processes tied to group membership shaped how the pandemic is remembered.

7 Limitations and Future Directions

This chapter provides a comprehensive review of the evidence on the collective mental time travel (MTT) during the COVID-19 pandemic, however several limitations in the current status of the literature must be noted. First, the cross-national comparison of collective representations is often constrained by differences in language, media environments, and pandemic policy responses. While the themes of infection, lockdown, and economic consequences showed convergence across nations, their meaning and emotional tone may vary by local context. For instance, themes such as "health system failure" or "social solidarity" may evoke different emotional and political interpretations depending on the country's institutional trust or government response (Fridman & Gensburger, 2023; Liu et al., 2005). Future studies should contextualize these collective representations with local sociopolitical data and media discourse analyses to better understand how meaning is co-constructed within groups.

It is also important to note that pandemic severity and governmental stringency have been shown to shape the content of collective representations. However, the mechanisms underlying the effects of these structural conditions on collective mental time travel might involve media mediation. In countries with high governmental stringency, themes of health and social solidarity were more frequent, suggesting the presence of media narratives designed to bolster public compliance and morale (e.g., Öner & Gülgöz, 2023). Conversely, in countries with lower stringency or institutional trust, future-oriented themes such as fear of a second wave or renewed lockdowns were more prominent, possibly reflecting media-driven uncertainty or political contestation (Liu et al., 2021). However, despite the centrality of media, most existing research on collective MTT during the pandemic relies on self-reported memories and expectations without directly assessing the information environments that participants were exposed to. Future studies should integrate content analyses of media coverage with individual-level cognitive data to more systematically examine how media narratives influence collective memory and prospection.

Third, the role of age, gender, and developmental stage in shaping collective MTT remains underexplored. While initial findings suggest generational differences in the emotional tone and specificity of collective events (e.g., Hjuler et al., 2025), systematic comparisons across age groups are scarce. Given that age-related changes in autobiographical memory influence phenomenological experience and future thinking (Cole et al., 2016), it is likely that similar patterns emerge in collective cognition.

Additionally, although the chapter discusses individual differences such as personality and affective traits, the intersectionality of these with group-level identities (e.g., political ideology, national identity) has yet to be examined in detail. For example, how do high trait anxiety and strong national identification interact to shape the valence or vividness of future-oriented collective representations?

Finally, the temporal boundaries of collective memory remain fluid and susceptible to change through competing events and societal shifts. As seen in the transition from pandemic-related concerns to the salience of war, politics, and environmental crises in later phases (Watson et al., 2025), future research should adopt a dynamic lens to study how collective memory is overwritten, recontextualized, or forgotten in light of emerging global events (Schuman & Rodgers, 2004; Hirst & Coman, 2018).

8 Conclusion

In this chapter, we discussed collective mental time travel during the COVID-19 pandemic, emphasizing the interconnected roles of collective memory and collective future thinking. Findings demonstrated that the pandemic served as a shared global event, deeply embedded in collective representations across both national and global contexts. Consistent themes, such as lockdowns, infections, and economic impacts, emerged internationally, underscoring how significant emotional and sociopolitical events become central elements of collective memory. Phenomenological features of pandemic memories, including emotional intensity and vividness, further highlighted how shared events are not only remembered but emotionally experienced by communities.

When we look forward, we believe future research should focus more on how media representation and political communication dynamically influence public reactions, memories, and expectations surrounding crises. It will be crucial to explore how simultaneous sociopolitical movements occurring at the local level during the pandemic (i.e., the Black Lives Matter protests) interact with and potentially modify collective mental time travel. Investigating these interrelations can expand our understanding regarding the broader mechanisms underlying collective memory formation and evolution over time. Such insights are essential, as understanding how societies collectively encode, recall, and anticipate critical events can enhance community resilience, inform strategic public communication during crises, and improve preparedness for future global challenges.

References

Abel, M., Umanath, S., & Barzykowski, K. (2022). Decades later: What World War II events are remembered as the most important ones? Implications of data collected in 12 countries. In H. L. Roediger III & J. V. Wertsch (Eds.), *Constructing national identity: Conflicting memories and narratives* (pp. 190–208). Oxford University Press.

Abel, M., Umanath, S., Fairfield, B., Takahashi, M., Roediger, H. L., III, & Wertsch, J. V. (2019). Collective memories across 11 nations for World War II: Similarities and differences regarding the most important events. *Journal of Applied Research in Memory and Cognition, 8*(2), 178–188. https://doi.org/10.1037/h0101836

Addis, D. R., Wong, A. T., & Schacter, D. L. (2007). Remembering the past and imagining the future: Common and distinct neural substrates during event construction and elaboration. *Neuropsychologia, 45*(7), 1363–1377. https://doi.org/10.1016/j.neuropsychologia.2006.10.016

Addis, D. R., Pan, L., Vu, M. A., Laiser, N., & Schacter, D. L. (2009). Constructive episodic simulation of the future and the past: Distinct subsystems of a core brain network mediate imagining and remembering. *Neuropsychologia, 47*(11), 2222–2238. https://doi.org/10.1016/j.neuropsychologia.2008.10.026

Ahn, J. N., Hu, D., & Vega, M. (2021). Changing pace: Using implementation intentions to enhance social distancing behavior during the COVID-19 pandemic. *Journal of Experimental Psychology: Applied, 27*(4), 762–772. https://doi.org/10.1037/xap0000385

Aizpurua, A., Migueles, M., & Aranberri, A. (2021). Prospective memory and positivity bias in the COVID-19 health crisis: The effects of aging. *Frontiers in Psychology, 12*, 666977. https://doi.org/10.3389/fpsyg.2021.666977

Anderson, R. J. (2012). Imagining novel futures: The roles of event plausibility and familiarity. *Memory, 20*(5), 443–451. https://doi.org/10.1080/09658211.2012.677450

Assmann, J., & Czaplicka, J. (1995). Collective memory and cultural identity. *New German Critique, 65*, 125–133. https://doi.org/10.2307/488538

Baldassare, M., & Katz, C. (1992). The personal threat of environmental problems as predictor of environmental practices. *Environment and Behavior, 24*(5), 602–616. https://doi.org/10.1177/0013916592245002

Bavel, J. J. V., Baicker, K., Boggio, P. S., Capraro, V., Cichocka, A., Cikara, M., et al. (2020). Using social and behavioural science to support COVID-19 pandemic response. *Nature Human Behaviour, 4*(5), 460–471. https://doi.org/10.1038/s41562-020-0884-z

Berntsen, D. (2021). Involuntary autobiographical memories and their relation to other forms of spontaneous thoughts. *Philosophical Transactions of the Royal Society B, 376*(1817), 20190693. https://doi.org/10.1098/rstb.2019.0693

Bertolotti, M., & Catellani, P. (2023). Counterfactual thinking as a prebunking strategy to contrast misinformation on COVID-19. *Journal of Experimental Social Psychology, 104*, 104404. https://doi.org/10.1016/j.jesp.2022.104404

Brown, N. R. (2021). The possible effects of the COVID-19 pandemic on the contents and organization of autobiographical memory: A transition-theory perspective. *Cognition, 212*, 104694. https://doi.org/10.1016/j.cognition.2021.104694

Burnett, H. J., Wang, Q., Zaragoza Scherman, A., & Hirst, W. (2023). Imagining the collective future in a global crisis: The impact of the COVID-19 pandemic on collective future thoughts across cultures. *Frontiers in Psychology, 14*, 1173962. https://doi.org/10.3389/fpsyg.2023.1173962

Byrne, R. M. (2016). Counterfactual thought. *Annual Review of Psychology, 67*, 135–157. https://doi.org/10.1146/annurev-psych-122414-033249

Cole, S. N., Markostamou, I., Watson, L. A., Barzykowski, K., Ergen, İ., Taylor, A., & Öner, S. (2023). Spontaneous past and future thinking about the COVID-19 pandemic across 14 countries: Effects of individual and country-level COVID-19 impact indicators. *Journal of Applied Research in Memory and Cognition, 12*(4), 502–512. https://doi.org/10.1037/mac0000071

Cole, S. N., Staugaard, S. R., & Berntsen, D. (2016). Inducing involuntary and voluntary mental time travel using a laboratory paradigm. *Memory & Cognition, 44*, 376–389. https://doi.org/10.3758/s13421-015-0564-9

Coman, A., Brown, A. D., Koppel, J., & Hirst, W. (2009). Collective memory from a psychological perspective. *International Journal of Politics, Culture, and Society IJPS, 22*(2), 125–141. https://doi.org/10.1007/s10767-009-9057-9

Coquet, M., Mercier, D., & Fleury-Bahi, G. (2019). Assessment of the exposure to coastal flood risk by inhabitants of French coasts: The effect of spatial optimism and temporal pessimism. *Ocean & Coastal Management, 177*, 139–147. https://doi.org/10.1016/j.ocecoaman.2019.05.004

CoronaVirus Resource Center. (2023). *Johns Hopkins University & Medicine*. https://coronavirus.jhu.edu/map.html

Cuc, A., Koppel, J., & Hirst, W. (2007). Silence in not golden: A case for socially-shared retrieval-induced forgetting. *Psychological Science, 18*, 727–733. https://doi.org/10.1111/j.1467-9280.2007.01967.x

Dalgleish, T., & Hitchcock, C. (2023). Transdiagnostic distortions in autobiographical memory recollection. *Nature Reviews Psychology, 2*(3), 166–182. https://doi.org/10.1038/s44159-023-00148-1

del Palacio-Gonzalez A., Berntsen, D., & Watson, L. A. (2017). Emotional intensity and emotion regulation in response to autobiographical memories during dysphoria. *Cognitive Therapy and Research, 41*(4), 530–542. https://doi.org/10.1007/s10608-017-9841-1

del Palacio-Gonzalez, A., & Berntsen, D. (2020). Memory-related emotion regulation and its relation to internalizing symptoms. *Cognitive Therapy and Research, 44*, 1162–1176. https://doi.org/10.1007/s10608-020-10137-w

Deng, W., Rosenblatt, A. K., Talhelm, T., & Putnam, A. L. (2023). People from the US and China think about their personal and collective future differently. *Memory & Cognition, 51*(1), 87–100. https://doi.org/10.3758/s13421-022-01344-9

De Saint-Laurent, C. (2018). Thinking through time: From collective memories to collective futures. In *Imagining collective futures* (pp. 59–81). Palgrave Macmillan.

Espinar-Herranz, K., Delgado-Lima, A. H., Villatoro, B. S., Garaboa, E. M., Gómez, V. S., Vides, L. G., Bouhaben, J., & Delgado-Losada, M. L. (2023). Memory, emotion, and quality of life in patients with long COVID-19. *Brain Sciences, 13*(12), 1670. https://doi.org/10.3390/brainsci13121670

Fleury-Bahi, G. (2008). Environmental risk: Perception and target with local versus global evaluation. *Psychological Reports, 102*(1), 185–193. https://doi.org/10.2466/pr0.102.1.185-193

Fridman, O., & Gensburger, S. (2023). Introduction: Unlocking memory studies: Understanding collective remembrance during and of Covid-19. In O. Fridman & S. Gensburger (Eds.), *The COVID-19 pandemic and memory: Remembrance, commemoration, and archiving in crisis* (pp. 1–20). Springer International. https://doi.org/10.1007/978-3-031-34597-5_1

García Ferrés, E. A., & DePalma, M. T. (2023). Counterfactual thinking may attenuate polarization of COVID-19 prevention behavior. *Social and Personality Psychology Compass, 17*(12), e12891. https://doi.org/10.1111/spc3.12891

Geiger, N., McLaughlin, B., & Velez, J. (2021). Not all boomers: Temporal orientation explains inter-and intra-cultural variability in the link between age and climate engagement. *Climatic Change, 166*(1), 1–20. https://doi.org/10.1007/s10584-021-03116-x

Gifford, R., Scannell, L., Kormos, C., Smolova, L., Biel, A., Boncu, S., et al. (2009). Temporal pessimism and spatial optimism in environmental assessments: An 18-nation study. *Journal of Environmental Psychology, 29*(1), 1–12. https://doi.org/10.1016/j.jenvp.2008.06.001

Gottlieb, M., & Dyer, S. (2020). Information and disinformation: Social media in the COVID-19 crisis. *Academic Emergency Medicine, 27*(7), 640–641. https://doi.org/10.1111/acem.14036

Hacıbektaşoğlu, D., Tekcan, A. İ., Bilge, R., & Boduroglu, A. (2023). The impact of group identity on the interaction between collective memory and collective future thinking negativity: Evidence from a Turkish sample. *Memory & Cognition, 51*(3), 752–772. https://doi.org/10.3758/s13421-022-01326-x

Halbwachs, H. (1925). *Les cadres sociaux de la mémoire*. Alcan.

Halbwachs, M. (1992). *On collective memory.* University of Chicago Press. https://doi.org/10.7208/chicago/9780226774497.001.0001

Harber, K. D., & Cohen, D. J. (2005). The emotional broadcaster theory of social sharing. *Journal of Language and Social Psychology, 24*, 382–400. https://doi.org/10.1177/0261927X05281426

Hassabis, D., Kumaran, D., Vann, S. D., & Maguire, E. A. (2007). Patients with hippocampal amnesia cannot imagine new experiences. *Proceedings of the National Academy of Sciences, 104*(5), 1726–1731. https://doi.org/10.1073/pnas.0610561104

Heux, L., Rathbone, C., Gensburger, S., Clifford, R., & Souchay, C. (2023). Collective memory and autobiographical memory: Perspectives from the humanities and cognitive sciences. *Wiley Interdisciplinary Reviews: Cognitive Science, 14*(3), e1635. https://doi.org/10.1002/wcs.1635

Hirst, W., & Coman, A. (2018). Building a collective memory: The case for collective forgetting. *Current Opinion in Psychology, 23*, 88–92. https://doi.org/10.1016/j.copsyc.2018.02.002

Hirst, W., & Echterhoff, G. (2012). Remembering in conversations: The social sharing and reshaping of memories. *Annual Review of Psychology, 63*(1), 55–79. https://doi.org/10.1146/annurev-psych-120710-100340

Hirst, W., & Manier, D. (2008). Towards a psychology of collective memory. *Memory, 16*(3), 183–200. https://doi.org/10.1080/09658210701811912

Hirst, W., Yamashiro, J. K., & Coman, A. (2018). Collective memory from a psychological perspective. *Trends in Cognitive Sciences, 22*(5), 438–451. https://doi.org/10.1016/j.tics.2018.02.010

Hiscott, J., Alexandri-di, M., Muscolini, M., Tassone, E., Palermo, E., Soultsioti, M., & Zevini, A. (2020). The global impact of the coronavirus pandemic. *Cytokine & Growth Factor Reviews, 53*, 1–9. https://doi.org/10.1016/j.cytogfr.2020.05.010

Hjuler, T. F., Lee, D., & Ghetti, S. (2025). Remembering history: Autobiographical memory for the COVID-19 pandemic lockdowns, psychological adjustment, and their relation over time. *Child Development, 96*, 55–70. https://doi.org/10.1111/cdev.14131

Holman, E. A., & Silver, R. C. (1998). Getting "stuck" in the past: Temporal orientation and coping with trauma. *Journal of Personality and Social Psychology, 74*(5), 1146–1163. https://doi.org/10.1037/0022-3514.74.5.1146

Kensinger, E. A. (2009). What factors need to be considered to understand emotional memories? *Emotion Review, 1*(2), 120–121. https://doi.org/10.1177/1754073908100436

Klein, O., Licata, L., & Pierucci, S. (2011). Does group identification facilitate or prevent collective guilt about past misdeeds? Resolving the paradox. *British Journal of Social Psychology, 50*(3), 563–572. https://doi.org/10.1111/j.2044-8309.2011.02028.x

Klein, S. B., Loftus, J., & Kihlstrom, J. F. (2002). Memory and temporal experience: The effects of episodic memory loss on an amnesic patient's ability to remember the past and imagine the future. *Social Cognition, 20*(5), 353–379. https://doi.org/10.1521/soco.20.5.353.21125

Levine, L. J., Whalen, C. K., Henker, B., & Jamner, L. D. (2005). Looking back on September 11, 2001: Appraised impact and memory for emotions in adolescents and adults. *Journal of Adolescent Research, 20*(4), 497–523. https://doi.org/10.1177/0743558405274893

Liu, J. H., Goldstein-Hawes, R., Hilton, D., et al. (2005). Social representations of events and peoples in world history across 12 cultures. *Journal of Cross-Cultural Psychology, 36*(2), 171–191. https://doi.org/10.1177/0022022104272900

Liu, J. H., Zhang, R. J., Vilar, R., Milojev, P., Hakim, M. A., de Zúñiga, H. G., Schumann, S., & Páez, D. (2021). A typology of mass personal information seeking repertoires (MISR): Global implications for political participation and subjective well-being. *New Media & Society, 23*(9), 2729–2753. https://doi.org/10.1177/1461444820932556

Luminet, O., IV, Bouts, P., Delie, F., Manstead, A. S., & Rimé, B. (2000). Social sharing of emotion following exposure to a negatively valenced situation. *Cognition & Emotion, 14*(5), 661–688. https://doi.org/10.1080/02699930050117666

Mather, M., & Carstensen, L. L. (2005). Aging and motivated cognition: The positivity effect in attention and memory. *Trends in Cognitive Sciences, 9*(10), 496–502. https://doi.org/10.1016/j.tics.2005.08.005

McBride, O., Butter, S., Murphy, J., Shevlin, M., Hartman, T. K., Hyland, P., McKay, R., Bennett, K. M., Gibson-Miller, J., Levita, L., Mason, L., Martinez, A. P., Stocks, T. V., Vallières, F., Karatzias, T., Valiente, C., Vazquez, C., & Bentall, R. P. (2021). Context, design and conduct of the longitudinal COVID-19 psychological research consortium study-wave 3. *International Journal of Methods in Psychiatric Research, 30*(3), e1880. https://doi.org/10.1002/mpr.1880

Meade, M. L., & Roediger, H. L. (2002). Explorations in the social contagion of memory. *Memory & Cognition, 30*(7), 995–1009. https://doi.org/10.3758/BF03194318

Merck, C., Topçu, M. N., & Hirst, W. (2016). Collective mental time travel: Creating a shared future through our shared past. *Memory Studies, 9*(3), 284–294. https://doi.org/10.1177/1750698016645236

Mert, N., Hou, Y., & Wang, Q. (2023). What lies ahead of us? Collective future thinking in Turkish, Chinese, and American adults. *Memory & Cognition, 51*(3), 773–790. https://doi.org/10.3758/s13421-022-01321-2

Michaelian, K., & Sutton, J. (2019). Collective mental time travel: Remembering the past and imagining the future together. *Synthese, 196*(12), 4933–4960. https://doi.org/10.1007/s11229-017-1449-1

Musson, C. (1974). Local attitudes to population growth in South Buckinghamshire. In H. B. Perry (Ed.), *Population and its problems: A plain man's guide* (pp. 392–393). Clarendon Press.

Nuttin, J. (1985). *Future time perspective and motivation*. Leuven University Press and Lawrence Erlbaum Associates.

Olick, J. K. (1999). Collective memory: The two cultures. *Sociological Theory, 17*(3), 333–348. https://doi.org/10.1111/0735-2751.00083

Öner, S., & Gülgöz, S. (2020). Representing the collective past: Public event memories and future simulations in Turkey. *Memory, 28*. (3), 386–398. https://doi.org/10.1080/09658211.2020.1727520

Öner, S., & Gülgöz, S. (2023). Lack of bump in public events when recent events prevail. *Memory Studies, 16*(5), 1360–1382. https://doi.org/10.1177/17506980221122149

Öner, S., Szpunar, K., Watson, L. A., & Cole, S. (2025). Spatial optimism in individuals' future thinking about the COVID-19 pandemic. *Applied Cognitive Psychology, 39*, e70080. https://doi.org/10.1002/acp.70080

Öner, S., Watson, L. A., Adıgüzel, Z., Ergen, İ., Bilgin, E., Curci, A., et al. (2023). Collective remembering and future forecasting during the COVID-19 pandemic: How the impact of COVID-19 affected the themes and phenomenology of global and national memories across 15 countries. *Memory & Cognition, 51*(3), 729–751. https://doi.org/10.3758/s13421-022-01329-8

Peetz, J., & Wohl, M. J. (2019). Perceiving time through group-based glasses: Collective temporal orientation. *British Journal of Social Psychology, 58*(3), 609–629. https://doi.org/10.1111/bjso.12291

Pennebaker, J. W., Páez, D., & Deschamps, J. C. (2006). The social psychology of history: Defining the most important events of the last 10, 100, and 1000 years. *Psicología Política, 32*, 15–32.

Qin, J., Mitchell, K. J., Johnson, M. K., Krystal, J. H., Southwick, S. M., Rasmusson, A. M., & Allen, E. S. (2003). Reactions to and memories for the September 11, 2001 terrorist attacks in adults with posttraumatic stress disorder. *Applied Cognitive Psychology: The Official Journal of the Society for Applied Research in Memory and Cognition, 17*(9), 1081–1097. https://doi.org/10.1002/acp.987

Richmond, B., Sharpe, L., & Menzies, R. E. (2022). Are fear campaigns effective for increasing adherence to COVID-Related Mitigation Measures? *International Journal of Behavioral Medicine, 30*(5), 714–730. https://doi.org/10.1007/s12529-022-10137-y

Rimé, B., Finkenauer, C., Luminet, O., Zech, E., & Philippot, P. (1998). Social sharing of emotion: New evidence and new questions. In W. Stroebe & M. Hewstone (Eds.), *European review of social psychology* (Vol. 9, pp. 145–189). Wiley.

Roese, N. J. (1994). The functional basis of counterfactual thinking. *Journal of Personality and Social Psychology, 66*(5), 805–818. https://doi.org/10.1037/0022-3514.66.5.805

Roese, N. J., & Olson, J. M. (1997). Counterfactual thinking: The intersection of affect and function. *Advances in Experimental Social Psychology, 29*, 1–59. https://doi.org/10.1016/S0065-2601(08)60015-5

Sani, F., Bowe, M., Herrera, M., Manna, C., Cossa, T., Miao, X., et al. (2007). Perceived collective continuity: Seeing groups as entities that move through time. *European Journal of Social Psychology, 37*(6), 1118–1134. https://doi.org/10.1002/ejsp.430

Schacter, D. L., & Addis, D. R. (2007). The cognitive neuroscience of constructive memory: Remembering the past and imagining the future. *Philosophical Transactions of the Royal Society B: Biological Sciences, 362*(1481), 773–786. https://doi.org/10.1098/rstb.2007.2087

Schuman, H., & Rodgers, W. L. (2004). Cohorts, chronology, and collective memories. *Public Opinion Quarterly, 68*(2), 217–254. https://doi.org/10.1093/poq/nfh012

Seifert, M., & Somasundaram, J. (2025). COVID-19 death toll predictions show that triggering counterfactual thinking deteriorates judgmental performance. *Communications Medicine, 5*(1), 1–8. https://doi.org/10.1038/s43856-025-00751-8

Shan, Y., Rubin, D. C., & Berntsen, D. (2025). Involuntary autobiographical memoriesas a transdiagnostic factor in mental disorders. *Clinical Psychology Review, 116*, 102545. https://doi.org/10.1016/j.cpr.2025.102545

Shipp, A. J., Edwards, J. R., & Lambert, L. S. (2009). Conceptualization and measurement of temporal focus: The subjective experience of the past, present, and future. *Organizational Behavior and Human Decision Processes, 110*(1), 1–22. https://doi.org/10.1016/j.obhdp.2009.05.001

Shrikanth, S., & Szpunar, K. K. (2021). The good old days and the bad old days: Evidence for a valence-based dissociation between personal and public memory. *Memory, 29*(2), 180–192. https://doi.org/10.1080/09658211.2020.1871024

Shrikanth, S., Szpunar, P. M., & Szpunar, K. K. (2018). Staying positive in a dystopian future: A novel dissociation between personal and collective cognition. *Journal of Experimental Psychology: General, 147*(8), 1200–1210. https://doi.org/10.1037/xge0000421

Skowronski, J. J., Walker, W. R., Henderson, D. X., & Bond, G. D. (2014). The fading affect bias: Its history, its implications, and its future. *Advances in Experimental Social Psychology, 49*, 163–218. https://doi.org/10.1016/B978-0-12-800052-6.00003-2

Soyaltin-Colella, D., & Sert, D. (2024). The strategic use of narratives and governance of the COVID-19 pandemic in major autocratisers in europe. *Journal of Balkan and Near Eastern Studies, 26*(4), 565–583. https://doi.org/10.1080/19448953.2024.2307820

Stone, C. B., & Jay, A. C. (2019). From the individual to the collective: The emergence of a psychological approach to collective memory. *Applied Cognitive Psychology, 33*(4), 504–515. https://doi.org/10.1002/acp.3564

Szpunar, K. K., Spreng, R. N., & Schacter, D. L. (2014). A taxonomy of prospection: Introducing an organizational framework for future-oriented cognition. *Proceedings of the National Academy of Sciences, 111*(52), 18414–18421. https://doi.org/10.1073/pnas.1417144111

Szpunar, P. M., & Szpunar, K. K. (2016). Collective future thought: Concept, function, and implications for collective memory studies. *Memory Studies, 9*(4), 376–389. i. https://doi.org/10.1177/1750698015615660

Topçu, M. N., & Hirst, W. (2020). Remembering a nation's past to imagine its future: The role of event specificity, phenomenology, valence, and perceived agency. *Journal of Experimental Psychology: Learning, Memory, and Cognition, 46*(3), 563–579. https://doi.org/10.1037/xlm0000746

Topçu, M. N., & Hirst, W. (2022). Collective mental time travel: Current research and future directions. *Progress in Brain Research, 274*(1), 71–97. https://doi.org/10.1016/bs.pbr.2022.06.002

Topçu, M. N., & Hirst, W. (2024). When the personal and the collective intersects: Memory, future thinking, and perceived agency during the COVID-19 pandemic. *Journal of Experimental Psychology: General, 153*(9), 2258–2278. https://doi.org/10.1037/xge0001624

Tulving, E. (1985). Memory and consciousness. *Canadian Psychology/Psychologie Canadienne, 26*(1), 1–12. https://doi.org/10.1037/h0080017

Tulving, E. (2002). Episodic memory: From mind to brain. *Annual Review of Psychology, 53*(1), 1–25. https://doi.org/10.1146/annurev.psych.53.100901.135114

Tversky, A., & Kahneman, D. (1974). Judgment under uncertainty: Heuristics and biases: Biases in judgments reveal some heuristics of thinking under uncertainty. *Science, 185*(4157), 1124–1131. https://doi.org/10.1126/science.185.4157.1124

Tversky, A., & Koehler, D. J. (1994). Support theory: A nonextensional representation of subjective probability. *Psychological Review, 101*(4), 547–567. https://doi.org/10.1037/0033-295X.101.4.547

Tvinnereim, E., Lægreid, O. M., Liu, X., Shaw, D., Borick, C., & Lachapelle, E. (2020). Climate change risk perceptions and the problem of scale: Evidence from cross-national survey experiments. *Environmental Politics, 29*(7), 1178–1198. https://doi.org/10.1080/09644016.2019.1708538

Watson, L. A., Berntsen, D., Kuyken, W., & Watkins, E. R. (2012). The characteristics of involuntary and voluntary autobiographical memories in depressed and never depressed individuals. *Consciousness and Cognition, 21*(3), 1382–1392. https://doi.org/10.1016/j.concog.2012.06.016

Watson, L. A., Öner, S., Cole, S., Barzykowski, K., Ilczuk, E., Bascón Díaz, M. J., ... Jin, Z. (2025). Are we already forgetting the Covid pandemic? Politics and Russia-Ukraine war replace Covid-related collective event themes across early and late phases of the pandemic *Manuscript submitted for publication.*

World Health Organization. (2020). *WHO director-general's opening remarks at the media briefing on COVID-19* [Press Briefing]. https://www.who.int/director-general/speeches/detail/who-director-general-s-opening-remarks-at-the-media-briefing-on-covid-19%2D%2D-11-march-2020

Yamashiro, J. K., & Roediger, H. L., III. (2019). How we have fallen: Implicit trajectories in collective temporal thought. *Memory, 27*(8), 1158–1166. https://doi.org/10.1080/09658211.2019.1635161

Yamashiro, J. K., Van Engen, A., & Roediger, H. L., III. (2022). American origins: Political and religious divides in US collective memory. *Memory Studies, 15*(1), 84–101. https://doi.org/10.1177/1750698019856065

Zaragoza Scherman, A., Burnett, H. J., Wang, Q., & Hirst, W. (2022). Remembering and imagining the pandemic: A cross-national study of the effects of age and country on collective memory and future thinking. *Frontiers in Psychology, 13*, 998121. https://doi.org/10.3389/fpsyg.2022.998121

Zhang, W., Gao, F., Gross, J., Shrum, L. J., & Hayne, H. (2021). How does social distancing during COVID-19 affect negative moods and memory? *Memory, 29*(1), 90–97. https://doi.org/10.1080/09658211.2020.1857774

Zimbardo, P. G., & Boyd, J. N. (1999). Putting Time in Perspective: A valid, reliable individual-differences metric. *Journal of Personality and Social Psychology, 77*(6), 1271–1288. https://doi.org/10.1037/0022-3514.77.6.1271

Involuntary Memories and Future Thoughts During the Covid-19 Pandemic

Mélissa C. Allé and Dorthe Berntsen

Abstract The Covid-19 pandemic abruptly disrupted the lives of millions, leading to a profound and radical transformation in daily routines and lifestyles. This unprecedented situation influenced spontaneous cognition, particularly involuntary mental time travel—the involuntary retrieval of autobiographical memories or the spontaneous imagination of future events. In this chapter, we first present three studies that explored involuntary mental time travel during the Covid-19 pandemic, examining its relationship with the pandemic context, lockdown conditions, and individual mental health. We then discuss the underlying mechanisms that may explain the observed increase in involuntary autobiographical memories and future thoughts during the Covid-19 pandemic, including changes in the availability of attentional resources, changes in life circumstances, and growing personal concerns. Finally, we examine the relationship between the increase in involuntary autobiographical memories and future thoughts—sometimes described as intrusive—and poorer mental health in individuals, highlighting potential traumatic stress reactions triggered by the Covid-19 crisis.

Keywords Involuntary retrieval · Future projections · Covid-19 · Living conditions · Mental health

M. C. Allé (✉)
Univ. Lille, CNRS, UMR 9193 – SCALab – Sciences Cognitives et Sciences Affectives, Lille, France
e-mail: melissa.alle@univ-lille.fr

D. Berntsen
Center on Autobiographical Memory Research, Department of Psychology and Behavioural Sciences, Aarhus University, Aarhus, Denmark
e-mail: dorthe@psy.au.dk

M. Migueles Seco, A. Aizpurua Sanz (eds.), *Autobiographical Memory and the Covid-19 Pandemic*, https://doi.org/10.1007/978-3-032-03895-1_4

1 Introduction

In 2020, as the world grappled with the Covid-19 pandemic, two major life disruptions emerged as a consequence of this global crisis. First, the pandemic itself created an unprecedented context of collective stress, fear of illness or death for ourselves and our loved ones, and a high level of uncertainty (Ekingen et al., 2023; Perz et al., 2022; Yıldırım et al., 2023). Second, as a collective response to protect the population, drastic measures of social isolation were implemented, forcing millions of people into lockdown or confinement. These measures led to an abrupt and radical change in people's way of life, resulting in the loss of familiar routines and a complete upheaval of daily life.

The present chapter considers how this unique situation might modulate spontaneous cognition in individuals, in particular involuntary mental time travel, and how these changes relate to mental health. First, we provide a brief overview of the cognitive processes underlying involuntary memories and future thoughts. Second, we present three studies that investigated how the pandemic context, including social isolation and lockdown measures, influenced involuntary autobiographical memory and future thinking. Third, from an adaptive cognitive perspective, we explore the underlying mechanisms that may account for the observed increase in involuntary autobiographical memories and future thoughts during the Covid-19 pandemic. Last, we highlight how involuntary mental time travel can sometimes become dysfunctional and be associated with psychopathology. We then discuss how the Covid-19 pandemic may have triggered traumatic stress reactions, including a maladaptive increase in intrusive memories and future thoughts.

2 Involuntary Autobiographical Memories

Involuntary autobiographical memories refer to memories for personal events that come to mind spontaneously with no preceding conscious attempt at retrieval (Berntsen, 1996). They simply "pop up" during our activities in daily life. Contrary to memories voluntarily and deliberately recalled based on strategic search into memory sustained by cognitive effort, involuntary memories arise from automatic and associative processes (Berntsen, 2010). Unlike voluntary autobiographical memories, involuntary memories more often involve specific episodes and tend to have a greater impact on mood at the moment of retrieval (Berntsen & Hall, 2004, see Berntsen, 2021, for a review).

Involuntary autobiographical memory represents a basic mode of remembering, and is considered a universal phenomenon, experienced by anyone (Berntsen, 2010). While involuntary memories occur frequently, at least as frequently than their voluntary counterpart (Rasmussen & Berntsen, 2011), their occurrence can vary both between different individuals and within the same individual over time. Involuntary autobiographical memories occur throughout individuals' lifespan and do not seem to decline with age, unlike other cognitive processes such as voluntary retrieval or daydreaming (Berntsen et al., 2015). However, they do vary according to individuals' mental health. Individuals with symptoms of depression, post-traumatic stress, anxiety, worry, psychosis, or rumination also report a higher frequency of involuntary autobiographical memories (Allé et al., 2018, 2023; Berntsen et al., 2015; Shan et al., 2025b). In everyday life, involuntary autobiographical memories, just as voluntary autobiographical memories are predominantly emotionally positive (Berntsen, 2009).

Imaginations of possible future events, referred to as episodic future thoughts (D'Argembeau & Van der Linden, 2006; Szpunar, 2010), also arise involuntarily during everyday life (Berntsen & Jacobsen, 2008; Finnbogadóttir & Berntsen, 2013). Hence, involuntary future events represent possible personal future events that come to mind involuntarily, with no preceding attempts at generating the imagined event. Involuntary future events appear to be as frequent as involuntary memories (Finnbogadóttir & Berntsen, 2013) and their occurrence conditions seems strikingly similar to those of involuntary memories (Berntsen & Jacobsen, 2008). Also, comparisons between involuntary versus voluntary future events largely mirror the differences observed between the involuntary and voluntary memories (Berntsen & Jacobsen, 2008; Finnbogadóttir & Berntsen, 2011).

As previously mentioned, the retrieval of involuntary memories relies on associative and automatic processes. Berntsen (2007) suggested that the involuntary retrieval of past events (and by extension the involuntary imagination of future events) is influenced by both motivational and environmental factors, which work together to optimize the probability that these memories are relevant to the situation in which they arise. Without this relevance, involuntary autobiographical memories would serve as mere disruptions of consciousness. She described two psychological levels that shape involuntary retrieval process: The immediate situation (e.g., the physical surroundings at the time of retrieval) and the current life situation (e.g., more extended concerns and goals; see Fig. 1 adapted from Berntsen, 2007,

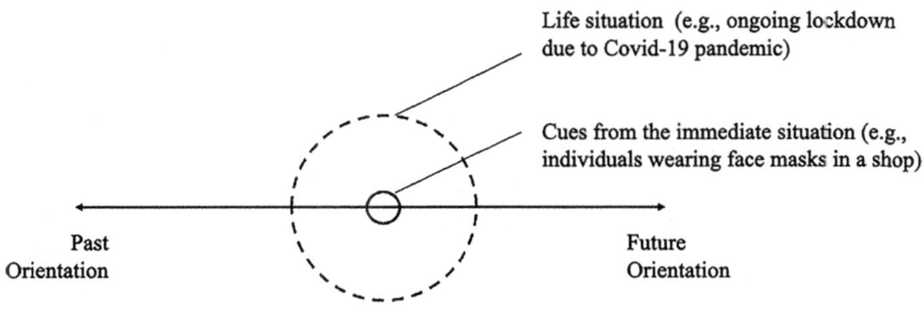

Fig. 1 Involuntary mental time travel arising from an interaction between cues in the immediate situation and the broader life context, including an individual's interests and concerns

Figure 2.1). First, the immediate situation refers to the specific moment when an involuntary memory occurs. Empirical research suggests that features of the immediate situational cues—such as an object, or a sensory impression—often trigger involuntary memories. More precisely, the involuntary recall of past events is facilitated when distinctive features in the current environment, either external elements (concrete objects, locations, people, or activities) or, less frequently, internal thoughts and feelings, align with the content of a past experience (see for review, Berntsen, 2021). The strength of these associations is further intensified by factors such as recency, uniqueness, and emotional significance of the remembered event (Berntsen, 2009). The distinctive feature overlap between the ongoing situation and the content of a memory represents a key mechanism of involuntary recall. Berntsen and Jacobsen (2008) also found that involuntary future events were triggered by situational cues in much the same ways as involuntary memories. These cues often involved overlapping situational features, such as specific objects, people, themes, or locations; indicating that the same contextual elements that bring past memories to mind can also prompt spontaneous thoughts about potential future events.

Second, the current life situation is a broader construct encompassing an individual's appraisal of their recent past and near future. It includes plans, goals, and ongoing cognitive processing of past events, whether stressful, pleasant or neutral. It was shown that long-term concerns (Johannessen & Berntsen, 2010) and recent emotionally impactful events (Berntsen, 2001) influence the content of involuntary memories. This influence could happen in at least two ways: The current life situation could prime clusters of autobiographical memories related to the same theme (e.g., my grandmother is sick) or it might exert its influence by unconsciously directing the person's attention towards theme-relevant cues (e.g., an ambulance in the street), which in turn may trigger an involuntary memory (e.g., when I visited her in the hospital) or future event (e.g., when I will visit her later this week) related to this concern. In other words, involuntary retrieval, or imagination, of personal events arise from the interaction between a cue in the immediate situation and the broader life situation of the person.

In addition, involuntary autobiographical memories and future projections are most likely to arise in situations where attention is diffuse, such as during monotonous tasks or in situations demanding minimal concentration (Berntsen, 1998; Schlagman & Kvavilashvili, 2008). When the immediate situation does not require significant attention, a person's thoughts may drift toward an appraisal of their current life situation. This, in turn, increases the likelihood of noticing a relevant cue—either in their thoughts or surroundings—that could trigger a memory. Also, concentrating on a specific task tends to limit associative activation due to cognitive control mechanisms that suppress knowledge units unrelated to the ongoing task (Mandler, 1994).

Following this view, overall, involuntary mental time travel (including both past and future involuntary mental events) is a specific form of spontaneous cognition that is particularly sensitive to and dependent both on the immediate context and the extended current life situation. Therefore, during times of crisis, such as the Covid-19 pandemic—which significantly disrupted individuals' daily lives, routines, and environments—this form of spontaneous cognition was particularly likely to be affected.

3 Research on Involuntary Autobiographical Memories in Times of Covid-19

In 2020, as the world was facing the Covid-19 pandemic and an unprecedented global lockdown, we investigated the impact of social isolation on cognitive functioning and mental health in individuals (Allé & Berntsen, 2021). Our study demonstrated that the enforced social isolation or confinement exacerbated cognitive problems among the general population in the USA. Among the various cognitive issues assessed—including intrusive thoughts, lack of impulse control, memory or concentration difficulties, intrusive future Covid-19-related thoughts, and voluntary future thoughts—intrusive future Covid-19-related thoughts were the most commonly reported experiences (see supplemental material, Table S2, Allé & Berntsen, 2021).

In this study, intrusive future Covid-19-related thoughts were defined as unwanted thoughts or images concerning potential stressful experiences in the future related to the coronavirus crisis. The occurrence of intrusive thoughts, in particular the future Covid-19-related intrusive thoughts, was significantly associated with higher levels of worries about Covid-19. Also, all types of intrusive thoughts (oriented toward past or future events) were negatively related to the size of living space per person during the lockdown and positively related to the number of persons within a household. And specifically, intrusive future Covid-19-related thoughts were also associated with greater life disruptions due to lockdown measures (see supplemental material, Table S3, Allé & Berntsen, 2021).

In a follow-up study by Shan et al. (2025a), researchers replicated and extended these findings in a sample of Chinese participants who answered a similar survey while they were isolated during the strict quarantine measures in China in the fall of 2022. During this period, China implemented some of the world's most stringent quarantine measures, requiring extended periods of confinement to enforce the nation's strict zero-COVID policy.

The findings were consistent with our earlier study (Allé & Berntsen, 2021), with intrusive thoughts about both the future and the past being the most frequently reported cognitive issues among Chinese participants (see supplemental material, Table S2, Shan et al., 2025a). In addition, unpublished correlation analyses from this study revealed that the occurrence of any type of intrusive thoughts were positively related with higher levels of social isolation, longer duration of confinement, higher worries about health, mental health, close ones, or financial situation, and elevated anxiety symptoms. In addition, the presence of intrusive future Covid-19-related thoughts was specifically associated with more extended social isolation, higher worries about health, mental health and financial situation, and increased anxiety symptoms.

During the Covid-19 crisis, Cole et al. (2022) also examined the extent to which participants reported experiencing spontaneous thoughts about personal pandemic-related events, both past and future, with regard to frequency, emotional valence, and emotional intensity. They conducted this investigation across multiple countries, considering collective protective measures and the severity of the pandemic (e.g., higher Covid-19 infection rates) in each location. Their findings revealed that governmental stringency measures predicted the frequency of spontaneous pandemic-related thoughts concerning both past and future events. Additionally, the severity of the pandemic predicted the frequency of spontaneous pandemic-related memories. Participants predominantly reported spontaneous thoughts about past and future pandemic-related events as negative, rating these thoughts as less positive than what is usually observed in the literature. Also, the authors did not observe the well-documented future positivity bias (D'Argembeau et al., 2011; Rasmussen & Berntsen, 2013) within the pandemic context.

These three studies demonstrated that the pandemic and the extraordinary circumstances of social isolation significantly impacted cognitive functioning, specifically increasing spontaneous cognitions. The occurrence of involuntary thoughts about past or future events related to Covid-19 was correlated with specific conditions of social isolation or lockdown and was also predicted by the severity of the pandemic. In the following section, we explore potential cognitive mechanisms that may explain an adaptive increase in involuntary mental time travel during the pandemic.

4 How the Pandemic Possibly Impacted Involuntary Autobiographical Memory Processes

As mentioned above, involuntary mental time travel is modulated by the availability of attentional resources, triggered by both external or internal cues, and influenced by current life concerns and emotional state. In this section, we examine how the pandemic period brought significant changes in these three factors, likely increasing the occurrence of involuntary memories and future thoughts related to the pandemic. Specifically, we examine how the pandemic (a) altered individuals' availability of attentional resources in two opposing ways—leading some to experience boredom while others faced cognitive overload and (b) reshaped both immediate situations and broader life concerns, fostering an excessive focus on Covid-19-related information, uncertainty about the future, and pandemic-related worries.

4.1 *Boredom Versus Overload*

Research on the impact of lockdowns on individuals has shown that two contrasting trajectories emerged in response to social isolation. First, for some individuals—such as the unemployed, students, those living alone, or the elderly—the changes induced by lockdown, along with self-isolation and quarantine, were associated with increased boredom, repetitive routines, and involuntary behaviors like overeating (Boylan et al., 2021; Droit-Volet et al., 2020; Latif & Karaman, 2021). The literature well documents that boredom and states of low mental load or diffuse attention are linked to a higher propensity for spontaneously generating these thoughts, particularly involuntary autobiographical memories (Berntsen, 1998, 2009) and involuntary future thoughts (Berntsen & Jacobsen, 2008; Berntsen, 2019).

The increased levels of boredom observed in some individuals during lockdown and confinement reflects a lower cognitive engagement in daily activities. This, in turn, leads to weaker inhibitory cognitive processes that typically regulate attentional focus, creating more mental space for associative processes. The decreased demand on attentional resources may have facilitated the emergence of involuntary memories and future-oriented thoughts, allowing them to surface more easily into consciousness without being inhibited.

On the other hand, it has been noted that for some individuals, lockdown was associated with a significantly higher mental load, in particular for working parents managing responsibilities from home (Delaney et al., 2023). For many women, this mental load became overwhelming due to the added burden of caregiving on top of domestic and paid work. In these situations, individuals were much less likely to experience boredom or states of diffuse attention, leaving little room for spontaneous thoughts to arise, including involuntary autobiographical memories or future projections.

4.2 *Overwhelming Covid-19-Related Information*

The studies mentioned above showed that individuals reported experiencing more intrusive thoughts related to both past and future events, particularly intrusive future thoughts concerning Covid-19 (Allé & Berntsen, 2021; Shan et al., 2025a). Throughout the pandemic, individuals were constantly exposed to a continuous flow of Covid-19-related information. This included public signage promoting hygiene practices and social distancing, daily media briefings, an overwhelming number of posts on social media as well as face masks, hand sanitizers, and Covid-19 test equipment. These pervasive environmental cues redefined individuals' immediate situation and likely acted as external triggers, directly eliciting involuntary thoughts related to the pandemic.

Beyond acting as direct triggers, the constant presence of epidemic reminders in everyday surroundings may have fostered a more subtle ongoing concern about the pandemic, heightening people's sensitivity to Covid-19-related thoughts. Consistent with these assumptions, Cole and colleagues further found that the increase in spontaneous thoughts about past and future Covid-19-related events was predicted by the stringency of government measures and the severity of the pandemic. These findings suggest that the overwhelming flow of information about the pandemic and its associated risks likely heightened individuals' concerns, reshaping their current life situation and attentional focus. In turn, this may have increased the frequency of spontaneous pandemic-related thoughts.

4.3 Uncertain Future and Lack of Control

The Covid-19 pandemic posed significant challenges to future thinking (Lalot et al., 2021). It required individuals to resist delay discounting—such as accepting present restrictions on social life and personal freedoms in anticipation of future improvements in public health—while also coping with profound uncertainty, given the constantly changing rules, infection rates, and behavioral recommendations (Rutter et al., 2020). In addition, the pandemic led to a diminished sense of control and necessitated self-sacrifice for the greater good of the community. Research has shown that maintaining a strong future consciousness and positive engagement with the future was beneficial for individuals. Those with higher future consciousness reported greater perceived well-being, reduced emotional blunting, and increased feelings of hope, although not necessarily reduced feelings of anger or anxiety (Lalot et al., 2021).

During the pandemic, high levels of uncertainty about the future were also associated with heightened expectations for what lies ahead, both on an individual and societal level (Yıldırım et al., 2023). Many people viewed the future with hope, anticipating an end to the difficult period of the pandemic and a return to normalcy, even if that normalcy differed from pre-pandemic life (Commodari & La Rosa, 2020). On a broader scale, the Covid-19 pandemic also prompted expectations for a different future in terms of socioeconomic development and sustainable policies, leading to increased discourse on reimagining political and developmental priorities (Gariboldi et al., 2021; Martín-Blanco et al., 2022).

Typically, people tend to recall their personal past and envision their personal future more positively than they do for collective past and future events (Shrikanth & Szpunar, 2021, but see Berntsen & Rubin, 2024). However, at the onset of the Covid-19 pandemic, Topçu and Hirst (2024) found that this positive bias, usually present in personal mental time travel compared to collective mental time travel, was observed only in memories, not in personal future projections. This shift likely occurred because the pandemic darkened how people envisioned their personal future. In the study by Topçu and Hirst, participants imagined more Covid-related events in their future than they recalled from their past. These findings suggest that the pandemic significantly clouded the typically optimistic outlook people have for their personal future, overshadowing it with uncertainty and concern for the upcoming close future.

Overall, the uncertainty and disruption caused by the pandemic made it challenging for individuals to make concrete plans and foresee the future during the Covid-19 crisis. In an effort to navigate these challenges, individuals' concerns and reflections likely became more future-oriented, though often overshadowed by uncertainty. Similar to the overwhelming influx of Covid-19-related information, persistent concerns about the future likely shaped the ongoing life situation, and thereby created a fertile ground for associative processes, facilitating the emergence of spontaneous, intrusive future thoughts related to the pandemic, as observed in the studies mentioned earlier (Allé & Berntsen, 2021; Cole et al., 2022; Shan et al., 2025a).

4.4 Covid-19-Related Worries

Building on the above mentioned future uncertainties and lack of control, worry is a cognitive process that involves anticipating potential threats and uncertainties, often focusing on future possibilities and the negative outcomes they might bring (Matthews & Funke, 2006). It is a natural response to situations perceived as dangerous or uncertain, as the mind attempts to prepare for and mitigate potential risks. Worry is strongly associated with poor mental health outcomes, and this association extends beyond the diagnosis of generalized anxiety disorder and clinical populations, as worry is also linked to poor mental health in nonclinical populations (Vîslă et al., 2022).

During the Covid-19 pandemic, a period marked by unprecedented uncertainty and widespread fear, worry levels significantly increased across the population (Varga et al., 2021). This was particularly evident during lockdowns and phases of social isolation. During the first lockdown, we observed that individuals with higher levels of Covid-19-related worry also reported experiencing more intrusive thoughts about potential future events related to the pandemic (Allé & Berntsen, 2021). In line with these findings, Cole et al. (2022) found that the severity of the pandemic predicted an increase in involuntary autobiographical memories. These results may indicate a condition characterized by heightened intrusive future projections about illness, death, or social and financial disruptions. In individuals with high levels of Covid-19-related worries, this maladaptive pattern of involuntary autobiographical memory could represent a key hallmark of the condition.

5 Dysfunctional Involuntary Autobiographical Memories

In everyday life, autobiographical memory is an adaptive process that serves various functions including identity support, social bonding, and guiding behavior (Pillemer, 2003; Rasmussen & Habermas, 2011). More specifically, the involuntary emergence of memories or potential future events is described as supporting individuals' adaptation to current situational demands (Berntsen, 2021). However, involuntary retrieval can also become dysfunctional and maladaptive. For example, intrusive memories are a key symptom of post-traumatic stress disorder (PTSD; *Diagnostic and Statistical Manual of Mental Disorders*, 2013). Berntsen (2021) suggests that in a dangerous environment and immediately following a traumatic event, frequent recall of the trauma may be functional by helping to prevent re-exposure to danger. However, as time passes and circumstances change, involuntary memories of the traumatic event should diminish. If they do not, they probably will be experienced as intrusive and distressing. Hence, intrusive memories are a dysfunctional side-effect of the normally functional involuntary memory mechanisms when applied to particularly stressful events, associated with intense emotional arousal and novelty, specifically in vulnerable populations (Berntsen, 2021).

In a clinical research, studies have primarily focused on involuntary memories that emerge following highly negative or stressful events, such as intrusive memories in PTSD. These memories are often distressing and are strongly associated with higher levels of depressive symptoms, as well as increased rumination and cognitive avoidance. Also, involuntary future projections are prevalent in clinical populations such as panic disorder and social anxiety disorder (see Clark, 1999, for review) and in relation to depression and bipolar disorder (Deeprose et al., 2011; Holmes et al., 2007). The investigation of suicidal imagery in depressed and previously suicidal patients showed that these images were vivid, persistent, and could be conceptualized as "flash-forwards" to the act of suicide (Holmes et al., 2007). In addition, Berntsen and Rubin (2015) demonstrated that involuntary intrusive images and thoughts of possible future events are also a significant aspect of PTSD phenomenology.

Beyond clinical investigations, several investigations demonstrated that a higher frequency of both voluntary and involuntary mental time travel is associated with increased emotional distress and negative thinking patterns, including anxiety, worry, depression, rumination, and hallucination-proneness (Allé et al., 2018, 2019; Berntsen et al., 2015). Overall, these findings highlight that a heightened frequency of involuntary autobiographical memories and future projections is not inherently beneficial but more likely to be linked to, or reflect, poor mental health outcomes (Shan et al., 2025a).

6 The Relationship Between Involuntary Mental Time Travel and Mental Health During the Covid-19 Pandemic

The uncertainty and quarantine measures associated with the Covid-19 pandemic have significantly impacted people's mental health. In 2020 alone, the pandemic contributed to a 27.6% increase of major depressive disorders and a 25.6% increase of anxiety disorders (COVID-19 Mental Disorders Collaborators, 2021). A review has shown that during the pandemic, the prevalence of depression reached 20%, anxiety 35%, and stress 53% in a combined study population of 113,285 individuals—markedly higher than pre-pandemic levels (Lakhan et al., 2020). In another review, Kan and colleagues reported that the prevalence of anxiety was 27.3% among the general population, while it was significantly higher at 39.6% among Covid-19 patients (Pashazadeh Kan et al., 2021). Hence, the pandemic and the resulting lockdowns had a profound effect on the mental health of populations, with certain subgroups being more at risk and requiring close monitoring. In particular, younger individuals and those with a history of mental illness would need higher attention to prevent resulting mental health issues from developing (Dadaczynski et al., 2022). As previously mentioned, individuals differ in the frequency with which they experience involuntary mental time travel in everyday life. Those with poorer mental health report more frequent involuntary autobiographical memories and future projections, regardless of context (Allé et al., 2018; Berntsen et al., 2015). During the Covid-19 pandemic, these individuals may be particularly vulnerable for the reasons outlined above, and should be more carefully monitored.

Relatedly, based on the studies reported above on involuntary mental time travel in times of Covid-19, we propose that the increased frequency of involuntary mental time travel—whether related to Covid-19 or not—among individuals currently experiencing social isolation may serve as a potential indicator of underlying mental health challenges. Intrusive memories and future thoughts represent central symptoms of PTSD and their higher frequency, and emotional intensity to a certain extent, are also related to a wide range of mental disorders symptoms, including depression, anxiety, psychosis (Allé et al., 2018; Berntsen et al., 2015). Hence, we suggest that the increased occurrence of intrusive thoughts observed, particularly those related to Covid-19 scenarios, and their associations with lockdown conditions or pandemic severity (Allé & Berntsen, 2021; Cole et al., 2022), may reflect a state of persistent traumatic stress induced by the Covid-19 crisis.

Although the Covid-19 pandemic does not align with traditional PTSD models or diagnostic criteria—which specify that a potentially traumatic event must involve exposure to the risk of imminent death, serious injury, or sexual violence (Criterion A of the DSM-5)—traumatic stress symptoms were widespread across the population in response to the global crisis. T*raumatic stress reactions* during the pandemic included intrusive re-experiencing and heightened arousal (Cooke et al., 2020; Zhang et al., 2020). Also, PTSD symptoms and intrusive traumatic memories have been particularly observed in Covid-19 survivors (Dinapoli et al., 2023).

Based on those observations and theoretical insights, several authors suggested that the Covid-19 pandemic and associated lockdown measures can be viewed as a traumatic period (Allé et al., 2020; Bridgland et al., 2021; Sanchez-Gomez et al., 2021). This perspective aligns with existing literature on social isolation and the effects of previous epidemics such as SARS and H1N1, which established a link between confinement and PTSD symptoms (Sprang & Silman, 2013). Unlike typical PTSD symptoms, Bridgland et al. (2021) proposed that people's traumatic stress reactions to the COVID-19 pandemic may be more closely related to factors such as the following: (1) concerns about the future rather than reflections on the past, (2) indirect exposure through media coverage rather than direct contact with the virus, and (3) stressful events like unemployment, isolation, and nonsudden illness or death, which do not meet Criterion A of the DSM-5 (i.e., actual or threatened death, injury, or sexual violation). These factors create a more nuanced and complex scenario compared to a "classic" traumatic event and invoke the concept of Continuous Traumatic Stress. This notion describes the experience and impact of living in contexts of persistent and realistic danger (Eagle & Kaminer, 2013) and seems particularly relevant in the context of a worldwide pandemic.

7 Conclusion

The worldwide Covid-19 pandemic led to an abrupt and radical change in people's way of life, reshaping their everyday cognition, in particular spontaneous cognition such as involuntary mental time travel. Our analyses point to three important factors whose combined effects may have generated a high prevalence of Covid-19 related involuntary memories and future thoughts: alterations in attentional resources (e.g., longer periods of boredom and lack of stimulation), alterations in the immediate situational context (e.g., an influx of Covid-19 related information), and alterations in the person's life situation (e.g., thoughts and worries about getting sick or losing one's job). For healthy and resourceful individuals, this increase in spontaneous thought may have been adaptive by spurring functional precautions and planning. However, for vulnerable individuals, the same factors may have generated maladaptive responses, such as symptoms of PTSD, anxiety and depression.

References

Allé, M. C., Berna, F., & Berntsen, D. (2018). Involuntary autobiographical memory and future thought predicting hallucination proneness. *Clinical Psychological Science, 6*(6), 1–8. https://doi.org/10.1177/2167702618785618

Allé, M. C., Berna, F., & Berntsen, D. (2019). Individuals with psychotic-like experiences exhibit enhanced involuntary autobiographical memories. *Psychiatry Research, 273*, 281–287. https://doi.org/10.1016/j.psychres.2019.01.050

Allé, M. C., Berna, F., Vidailhet, P., Giersch, A., & Mengin, A. C. (2020). Le confinement peut-il favoriser l'émergence de symptômes traumatiques ou psychotiques ? *Revue de Neuropsychologie, 12*(2), 196–203. https://doi.org/10.1684/nrp.2020.0570

Allé, M. C., & Berntsen, D. (2021). Self-isolation, psychotic symptoms and cognitive problems during the COVID-19 worldwide outbreak. *Psychiatry Research, 302*, 114015. https://doi.org/10.1016/j.psychres.2021.114015

Allé, M. C., Rubin, D. C., & Berntsen, D. (2023). Autobiographical memory and the self on the psychosis continuum: Investigating their relationship with positive- and negative-like symptoms. *Memory, 31*(4), 518–529. https://doi.org/10.1080/09658211.2023.2173236

Berntsen, D. (1996). Involuntary autobiographical memories. *Applied Cognitive Psychology, 10*(5), 435–454. https://doi.org/10.1002/(SICI)1099-0720(199610)10:5<435::AID-ACP408>3.0.CO;2-L

Berntsen, D. (1998). Voluntary and involuntary access to autobiographical memory. *Memory, 6*(2), 113–141. https://doi.org/10.1080/741942071

Berntsen, D. (2001). Involuntary memories of emotional events: Do memories of traumas and extremely happy events differ? *Applied Cognitive Psychology, 15*(7), S135–S158. https://doi.org/10.1002/acp.838

Berntsen, D. (2009). *Involuntary autobiographical memories: An introduction to the unbidden past.* Cambridge University Press.

Berntsen, D. (2010). The unbidden past: Involuntary autobiographical memories as a basic mode of remembering. *Current Directions in Psychological Science, 19*(3), 138–142. https://doi.org/10.1177/0963721410370301

Berntsen, D. (2007). Involuntary autobiographical memories: Speculations, findings and an attempt to integrate them. In J. H. Mace (Ed.), *Involuntary memory* (pp. 20–50). Blackwell.

Berntsen, D., & Rubin, D. C. (2015). Pretraumatic stress reactions in soldiers deployed to Afghanistan. *Clinical Psychological Science, 3*(5), 663–674.

Berntsen, D. (2019). Spontaneous future cognitions: An integrative review. *Psychological Research, 83*(4), 651–665. https://doi.org/10.1007/s00426-018-1127-z

Berntsen, D. (2021). Involuntary autobiographical memories and their relation to other forms of spontaneous thoughts. *Philosophical Transactions of the Royal Society of London. Series B, Biological Sciences, 376*(1817), 20190693. https://doi.org/10.1098/rstb.2019.0693

Berntsen, D., & Hall, N. M. (2004). The episodic nature of involuntary autobiographical memories. *Memory & Cognition, 32*(5), 789–803.

Berntsen, D., & Jacobsen, A. S. (2008). Involuntary (spontaneous) mental time travel into the past and future. *Consciousness and Cognition, 17*(4), 1093–1104. https://doi.org/10.1016/j.concog.2008.03.001

Berntsen, D., & Rubin, D. C. (2024). Collectives closer to the self are anticipated to have a brighter future: Self-enhancement in collective cognition. *Journal of Experimental Psychology: General, 153*(5), 1226–1235. https://doi.org/10.1037/xge0001550

Berntsen, D., Rubin, D. C., & Salgado, S. (2015). The frequency of involuntary autobiographical memories and future thoughts in relation to daydreaming, emotional distress, and age. *Consciousness and Cognition, 36*, 352–372. https://doi.org/10.1016/j.concog.2015.07.007

Boylan, J., Seli, P., Scholer, A. A., & Danckert, J. (2021). Boredom in the COVID-19 pandemic: Trait boredom proneness, the desire to act, and rule-breaking. *Personality and Individual Differences, 171*, 110387. https://doi.org/10.1016/j.paid.2020.110387

Bridgland, V. M. E., Moeck, E. K., Green, D. M., Swain, T. L., Nayda, D. M., Matson, L. A., Hutchison, N. P., & Takarangi, M. K. T. (2021). Why the COVID-19 pandemic is a traumatic stressor. *PLoS One, 16*(1), e0240146. https://doi.org/10.1371/journal.pone.0240146

Clark, D. M. (1999). Anxiety disorders: Why they persist and how to treat them. *Behaviour Research and Therapy, 37*(Suppl 1), 5–27. https://doi.org/10.1016/s0005-7967(99)00048-0

Cole, S., Markostamou, I., Watson, L., Barzykowski, K., Ergen, I., Taylor, A., & Oner, S. (2022). Spontaneous past and future thinking about the COVID-19 pandemic across 14 countries: Effects of individual and country-level COVID-19 impact indicators. *Journal of Applied Research in Memory and Cognition, 12*(4), 502–512. https://doi.org/10.1037/mac0000071

Commodari, E., & La Rosa, V. L. (2020). Adolescents in quarantine during COVID-19 pandemic in Italy: Perceived health risk, beliefs, psychological experiences and expectations for the future. *Frontiers in Psychology, 11*, 559951. https://doi.org/10.3389/fpsyg.2020.559951

Cooke, J. E., Eirich, R., Racine, N., & Madigan, S. (2020). Prevalence of posttraumatic and general psychological stress during COVID-19: A rapid review and meta-analysis. *Psychiatry Research, 292*, 113347. https://doi.org/10.1016/j.psychres.2020.113347

COVID-19 Mental Disorders Collaborators. (2021). Global prevalence and burden of depressive and anxiety disorders in 204 countries and territories in 2020 due to the COVID-19 pandemic. *Lancet, 398*(10312), 1700–1712. https://doi.org/10.1016/S0140-6736(21)02143-7

Dadaczynski, K., Okan, O., Messer, M., & Rathmann, K. (2022). University students' sense of coherence, future worries and mental health: Findings from the German COVID-HL-survey. *Health Promotion International, 37*(1), daab070. https://doi.org/10.1093/heapro/daab070

D'Argembeau, A., Renaud, O., & Van der Linden, M. (2011). Frequency, characteristics and functions of future-oriented thoughts in daily life. *Applied Cognitive Psychology, 25*(1), 96–103. https://doi.org/10.1002/acp.1647

D'Argembeau, A., & Van der Linden, M. (2006). Individual differences in the phenomenology of mental time travel: The effect of vivid visual imagery and emotion regulation strategies. *Consciousness and Cognition, 15*(2), 342–350. https://doi.org/10.1016/j.concog.2005.09.001

Deeprose, C., Malik, A., & Holmes, E. A. (2011). Measuring Intrusive Prospective Imagery using the Impact of Future Events Scale (IFES): Psychometric properties and relation to risk for Bipolar Disorder. *International Journal of Cognitive Therapy, 4*(2), 187–196. https://doi.org/10.1521/ijct.2011.4.2.187

Delaney, C., Bobek, A., & Clavero, S. (2023). "It was too much for me": Mental load, mothers, and working from home during the COVID-19 pandemic. *Frontiers in Psychology, 14*, 1208099. https://doi.org/10.3389/fpsyg.2023.1208099

Diagnostic and statistical manual of mental disorders: DSM-5™ (5th edn, pp. xliv, 947). (2013). American Psychiatric Publishing. https://doi.org/10.1176/appi.books.9780890425596

Dinapoli, L., Ferrarese, D., Belella, D., Carnevale, S., Camardese, G., Sani, G., & Chieffo, D. P. R. (2023). Psychological treatment of traumatic memories in COVID-19 survivors. *Clinical Psychology & Psychotherapy, 30*(1), 225–233.

Droit-Volet, S., Gil, S., Martinelli, N., Andant, N., Clinchamps, M., Parreira, L., Rouffiac, K., Dambrun, M., Huguet, P., Dubuis, B., Pereira, B., Network, C., Bouillon, J.-B., & Dutheil, F. (2020). Time and Covid-19 stress in the lockdown situation: Time free, «Dying» of boredom and sadness. *PLoS One, 15*(8), e0236465. https://doi.org/10.1371/journal.pone.0236465

Eagle, G., & Kaminer, D. (2013). Continuous traumatic stress: Expanding the lexicon of traumatic stress. *Peace and Conflict: Journal of Peace Psychology, 19*, 85. https://doi.org/10.1037/a0032485

Ekingen, E., Teleş, M., Yıldız, A., & Yıldırım, M. (2023). Mediating effect of work stress in the relationship between fear of COVID-19 and nurses' organizational and professional turnover intentions. *Archives of Psychiatric Nursing, 42*, 97–105. https://doi.org/10.1016/j.apnu.2022.12.027

Finnbogadóttir, H., & Berntsen, D. (2011). Involuntary and voluntary mental time travel in high and low worriers. *Memory, 19*(6), 625–640. https://doi.org/10.1080/09658211.2011.595722

Finnbogadóttir, H., & Berntsen, D. (2013). Involuntary future projections are as frequent as involuntary memories, but more positive. *Consciousness and Cognition, 22*(1), 272–280. https://doi.org/10.1016/j.concog.2012.06.014

Gariboldi, M. I., Lin, V., Bland, J., Auplish, M., & Cawthorne, A. (2021). Foresight in the time of COVID-19. *The Lancet Regional Health – Western Pacific, 6*, 100049. https://doi.org/10.1016/j.lanwpc.2020.100049

Holmes, E. A., Crane, C., Fennell, M. J. V., & Williams, J. M. G. (2007). Imagery about suicide in depression—"Flash-forwards"? *Journal of Behavior Therapy and Experimental Psychiatry, 38*(4), 423–434. https://doi.org/10.1016/j.jbtep.2007.10.004

Johannessen, K. B., & Berntsen, D. (2010). Current concerns in involuntary and voluntary autobiographical memories. *Consciousness and Cognition, 19*(4), 847–860. https://doi.org/10.1016/j.concog.2010.01.009

Lakhan, R., Agrawal, A., & Sharma, M. (2020). Prevalence of depression, anxiety, and stress during COVID-19 pandemic. *Journal of Neurosciences in Rural Practice, 11*(4), 519–525. https://doi.org/10.1055/s-0040-1716442

Lalot, F., Abrams, D., Ahvenharju, S., & Minkkinen, M. (2021). Being future-conscious during a global crisis: The protective effect of heightened Futures Consciousness in the COVID-19 pandemic. *Personality and Individual Differences, 178*, 110862. https://doi.org/10.1016/j.paid.2021.110862

Latif, H., & Karaman, E. (2021). COVID-19: Boredom in the family. *The Family Journal, 29*(2), 147–152. https://doi.org/10.1177/1066480720986496

Mandler, G. (1994). Hypermnesia, incubation, and mind-popping: On remembering without really trying. In C. Umiltà & M. Moscovitch (Eds.), *Attention and performance XV: Conscious and nonconscious information processing* (pp. 3–33). MIT Press.

Martín-Blanco, C., Zamorano, M., Lizárraga, C., & Molina-Moreno, V. (2022). The impact of COVID-19 on the sustainable development goals: Achievements and expectations. *International Journal of Environmental Research and Public Health, 19*(23), 23. https://doi.org/10.3390/ijerph192316266

Matthews, G., & Funke, G. J. (2006). Worry and information-processing. In A. Wells & G. Davey (Eds.), *Worry and its psychological disorders* (pp. 51–67). Wiley. https://doi.org/10.1002/9780470713143.ch4

Pashazadeh Kan, F., Raoofi, S., Rafiei, S., Khani, S., Hosseinifard, H., Tajik, F., Raoofi, N., Ahmadi, S., Aghalou, S., Torabi, F., Dehnad, A., Rezaei, S., Hosseinipalangi, Z., & Ghashghaee, A. (2021). A systematic review of the prevalence of anxiety among the general population during the COVID-19 pandemic. *Journal of Affective Disorders, 293*, 391–398. https://doi.org/10.1016/j.jad.2021.06.073

Perz, C. A., Lang, B. A., & Harrington, R. (2022). Validation of the fear of COVID-19 scale in a US college sample. *International Journal of Mental Health and Addiction, 20*(1), 273–283. https://doi.org/10.1007/s11469-020-00356-3

Pillemer, D. B. (2003). Directive functions of autobiographical memory: The guiding power of the specific episode. *Memory, 11*(2), 193–202. https://doi.org/10.1080/741938208

Rasmussen, A. S., & Berntsen, D. (2011). The unpredictable past: Spontaneous autobiographical memories outnumber autobiographical memories retrieved strategically. *Consciousness and Cognition, 20*(4), 1842–1846. https://doi.org/10.1016/j.concog.2011.07.010

Rasmussen, A. S., & Berntsen, D. (2013). The reality of the past versus the ideality of the future: Emotional valence and functional differences between past and future mental time travel. *Memory & Cognition, 41*(2), 187–200. https://doi.org/10.3758/s13421-012-0260-y

Rasmussen, A. S., & Habermas, T. (2011). Factor structure of overall autobiographical memory usage: The directive, self and social functions revisited. *Memory, 19*(6), 597–605. https://doi.org/10.1080/09658211.2011.592499

Rutter, H., Wolpert, M., & Greenhalgh, T. (2020). Managing uncertainty in the covid-19 era. *BMJ, 370*, m3349. https://doi.org/10.1136/bmj.m3349

Sanchez-Gomez, M., Giorgi, G., Finstad, G. L., Urbini, F., Foti, G., Mucci, N., Zaffina, S., & León-Perez, J. M. (2021). COVID-19 pandemic as a traumatic event and its associations with fear and mental health: A cognitive-activation approach. *International Journal of Environmental Research and Public Health, 18*(14), 7422. https://doi.org/10.3390/ijerph18147422

Schlagman, S., & Kvavilashvili, L. (2008). Involuntary autobiographical memories in and outside the laboratory: How different are they from voluntary autobiographical memories? *Memory & Cognition, 36*(5), 920–932.

Shan, Y., Berntsen, D., Jia, Y., Allé, M. C., & Rubin, D. C. (2025a). Exploring changes in consciousness, cognition, and anxiety amidst social isolation: China's strict Covid-19 measures as a case study. *Psychology of Consciousness: Theory, Research, and Practice.*

Shan, Y., Rubin, D. C., & Berntsen, D. (2025b). Involuntary autobiographical memories as a transdiagnostic factor in mental disorders. *Clinical Psychology Review, 116*, 102545. https://doi.org/10.1016/j.cpr.2025.102545

Shrikanth, S., & Szpunar, K. K. (2021). The good old days and the bad old days: Evidence for a valence-based dissociation between personal and public memory. *Memory, 29*(2), 180–192. https://doi.org/10.1080/09658211.2020.1871024

Sprang, G., & Silman, M. (2013). Posttraumatic stress disorder in parents and youth after health-related disasters. *Disaster Medicine and Public Health Preparedness, 7*(1), 105–110. https://doi.org/10.1017/dmp.2013.22

Szpunar, K. K. (2010). Episodic future thought: An emerging concept. *Perspectives on Psychological Science: A Journal of the Association for Psychological Science, 5*(2), 142–162. https://doi.org/10.1177/1745691610362350

Topçu, M. N., & Hirst, W. (2024). When the personal and the collective intersects: Memory, future thinking, and perceived agency during the COVID-19 pandemic. *Journal of Experimental Psychology. General*. https://doi.org/10.1037/xge0001624

Varga, T. V., Bu, F., Dissing, A. S., Elsenburg, L. K., Bustamante, J. J. H., Matta, J., van Zon, S. K. R., Brouwer, S., Bültmann, U., Fancourt, D., Hoeyer, K., Goldberg, M., Melchior, M., Strandberg-Larsen, K., Zins, M., Clotworthy, A., & Rod, N. H. (2021). Loneliness, worries, anxiety, and precautionary behaviours in response to the COVID-19 pandemic: A longitudinal analysis of 200,000 Western and Northern Europeans. *The Lancet Regional Health – Europe, 2*. https://doi.org/10.1016/j.lanepe.2020.100020

Vîslă, A., Stadelmann, C., Watkins, E., Zinbarg, R. E., & Flückiger, C. (2022). The relation between worry and mental health in nonclinical population and individuals with anxiety and depressive disorders: A meta-analysis. *Cognitive Therapy and Research, 46*(3), 480–501. https://doi.org/10.1007/s10608-021-10288-4

Yıldırım, M., Çağış, Z. G., & Williams, G. (2023). Fear of COVID-19, intolerance of uncertainty, psychological capital, and positive future expectations: Tests of mediating relationships with healthcare workers. *Archives of Psychiatric Nursing, 45*, 158–163. https://doi.org/10.1016/j.apnu.2023.06.016

Zhang, Q., Zheng, R., Fu, Y., Mu, Q., & Li, J. (2020). Post traumatic stress disorder and death anxiety symptoms during the coronavirus epidemic in 2019: A cross-sectional study based on the affected population, *21*, 1499. https://doi.org/10.21203/rs.3.rs-80116/v1

The Flashbulb-Like Nature of Memories for the Covid-19 Pandemic

Tiziana Lanciano and Antonietta Curci

Abstract Worldwide events such as the Covid-19 pandemic have the power to outline people's life pathways and memories for experiences, shaping both personal, collective, and historical narratives. Autobiographical memory is essential to enable people to approach future events with greater consciousness and learn from past circumstances. The unprecedented distinctiveness of the Covid-19 pandemic gave us the opportunity to explore the features of both collective and personal memory, such as Flashbulb Memories (FBMs). When a significant, impactful, and unexpected public event takes place, individuals tend to retain not only the details surrounding the event itself, but also irrelevant and peripheral details related to the context in which they received the news, such as date, location, informant, ongoing activity, or others present. In the realm of autobiographical memory, FBMs are determined by both emotional (i.e., intensity of emotional feeling, appraisal of the original event, importance-consequentiality) and social (social sharing, rumination, mass media) factors which give these memories high vividness, confidence, and consistency over time. The present chapter will overview the main empirical evidence about the features of FBM for public events such as the Covid-19 pandemic.

Keywords Flashbulb memories · Covid-19 · Pandemic · Autobiographical memories · Collective memories

1 Introduction

The present chapter explores how the unprecedented global experience of the Covid-19 pandemic has shaped the formation of photographic memories, called Flashbulb Memories (FBMs). The chapter begins with an overall key finding of

T. Lanciano (✉) · A. Curci
Department of Education, Psychology, Communication, University of Bari "Aldo Moro", Bari, Italy
e-mail: tiziana.lanciano@uniba.it; antonietta.curci@uniba.it

M. Migueles Seco, A. Aizpurua Sanz (eds.), *Autobiographical Memory and the Covid-19 Pandemic*, https://doi.org/10.1007/978-3-032-03895-1_5

studies on autobiographical memory of Covid-19, followed by a review of the limited empirical evidence on FBMs of Covid-19 events. Special focus is addressed to the novel feature of the pandemic compared to traditionally studied target events for FBM: Rather than being a single, episodic occurrence, the pandemic was prolonged and widespread, and this characteristic raises important implications for the formation of both personal and collective autobiographical memories. The chapter concludes by outlining the main models of FBMs, emphasizing the roles of emotional, cognitive, and social factors in their formation and maintenance.

The first case of Coronavirus 2019 (Covid-19) on 8th December 2019 in Wuhan has marked the beginning of the pandemic crisis with devastating worldwide effects. Human behaviors, lifestyles, and mental health issues were significantly altered; both directly (e.g., the individuals' continuous fear and worry of contagion) and indirectly (e.g., the government's restrictive measures and mitigation strategies). The hugeness of the pandemic and its effects on many psychological, social, political, and economic domains indicate that this has been—and is still being—a watershed event, very similar to a traumatic experience with significantly long-term effects (Adams & Kopelman, 2021; Castillo et al., 2024; Chamaa et al., 2021; Lanciano et al., 2020; Liu et al., 2020; Taurisano et al., 2022).

Maintaining distance has been the most frequently preventive strategy adopted for safeguarding people's physical health, and reducing infection rates and death (Yang et al., 2020). However, physical restrictions and limitations induced psychological and social isolation, leading to detrimental effects on people's mental well-being (Jacobson et al., 2020; Tull et al., 2020). Furthermore, extraordinary and uncommon daily routines during lockdowns would have diminished the chance to socially shared life events (Habermas, 2012; Hjuler et al., 2025; Pillemer & Kuwabara, 2012), which could have an impact on people's ability to recall and reflect on the past itself (Brown, 2021).

To a deep investigation of the short- and long-term Covid-19 effects, it seems noteworthy to examine the consequences in terms of past memories and future thoughts. In contrast to other public events that only affected a small number of people, Covid-19 was unique in that it has been a worldwide emergency. The Covid-19 pandemic's events are expected to have a lasting impression on history, extending beyond national boundaries and becoming ingrained in the collective memory of nations and the world, with a global importance (Ma & Wan, 2022). Covid-19-related memories have the potential to foster a feeling of self-continuity, direct behavior in the present and the future, and promote interpersonal relationships (Wolf & Nusser, 2022).

The unprecedented distinctiveness of the Covid-19 pandemic has created the best-case situation to explore the features of both collective and personal memory for pandemic-related events, such as FBMs, since all people around the world have experienced this unexpected and stressful event.

2 Remembering Covid-19 Pandemic

Memory may be seen as a useful tool for offering direction and comfort by generating associations from the present to the past, particularly during uncertain and crisis-filled periods (Liashenko et al., 2020; Ma & Wan, 2022). Worldwide events have the power to shape our life paths and hone our memories for experiences that will outline both our personal and historical narratives (Rouhani et al., 2023). Memory retention is essential to enable people to approach future issues with greater awareness and generate insights from past circumstances (Adams & Kopelman, 2021). By using the collective and personal memory of similar prior events, society can better prepare for and respond to a crisis, which has profound consequences for public health. Recollections of the SARS-CoV-2 pandemic will probably be important in future discussions about pandemic preparedness and the proper course of political actions (Sprengholz et al., 2023). Remembering exceptional and significant events, like a pandemic, is vital in the immediate and long-term, due to the unpredictability and dynamic nature of public events (Lutz et al., 2019). This is due to the possibility that an event will recur or endure over time, requiring ongoing awareness and reflection in order to adjust and manage the situation.

A growing number of studies (Brown, 2021; Cole et al., 2023; Öner et al., 2022; Rouhani et al., 2023; Rossi-Arnaud et al., 2023) have been investigating the content and the organization of autobiographical memories related to Covid-19 pandemic (i.e., lockdown, remote learning, past and future thoughts, emotions or contact tracking), and their impact on adjustment and general well-being. During the pandemic, Brown (2021) published one of the first papers about the autobiographical memories for Covid-19 events. The author made predictions about the possibility of adopting the *Transition Theory* Framework (Brown, 2016) to explain how memories of the Covid-19 pandemic differ from other historically important public events and how Covid-19 is likely to impact people's memories. Results of a recent study point in the same direction as these predictions. By gathering word-cued memories from the recent past, Ekinci and Brown (2024) discovered that participants routinely dated both Covid-related and unrelated memories using the pandemic as a temporal landmark, much more than other life periods. Similarly, in their 2024 study, Adıgüzel and colleagues showed that the COVID-19 pandemic has caused a breaking point in the individual memories' sequence, with memories functioning worse at the start of the pandemic than later.

The Covid-19 pandemic has also impacted on the themes reported in past memories and future thoughts. Öner et al. (2022) found that whereas future events (both national and international) contained themes of economy, a second wave and lockdown, past events (both national and international) reported topics of infections, lockdown, and politics. Through the same Öner et al.'s multi-country collecting data (2022), Cole et al. (2023) showed that the frequency of people's spontaneous thoughts linked to the pandemic was predicted both independently and combined by national and individual variables (i.e., stringency of measures and attention to

COVID-related information, respectively). Furthermore, by four studies across 11 countries, Sprengholz et al. (2023) showed that recollections of SARS-CoV-2 pandemic are strongly biased and distorted by people's current motivation, attitudes, and perceptions, such as their assessment of previous political conduct and their intentions for future behavior.

About the emotions of pandemic-related memories, Rouhani et al. (2023) examined the effects of emotions experienced and recalled throughout 2020 on the structure and content of long-term memory. The authors found that while lockdowns shortened recalled time, pandemic news were more memorable, unexpected, and unfavorable. The form and content of memories were further increased by the level of experienced emotions; specifically, negative emotions improved recall, while chronic emotional symptoms—such as depression and post-traumatic stress disorder—predicted more blunted memories. Additionally, Hjuler et al. (2025) tested age and gender differences in autobiographical memories of the Covid-19 lockdowns and explored whether the content of these memories impacted psychological adjustment over time. A sample of 247 Danish students was assessed three times (June 2020, January 2021, June 2021), and the results demonstrated that emotional valence of memories and their richness of details declined over time. Adolescent girls fared the poorest as psychological well-being weakened as time passed. Crucially, memories including factual information regarding Covid-19 and the lockdown, as well as greater levels of negative emotion, predicted lower psychological well-being over time. These findings highlighted features of autobiographical memory that may help to mitigate the detrimental effects of the lockdown.

3 FBMs: Definition and Characteristics

When a significant, upsetting, and unexpected public event takes place, it is observed that individuals tend to retain not only the details surrounding the event itself but also elements related to the context in which they received the news. This particular form of memory was initially introduced by Brown and Kulik in 1977, who named it "Flashbulb Memory" (FBM). The context in which individuals learn public news has traditionally been defined by the canonical categories: Where people were when they received the news, their ongoing activity at the time, the source of the news, the aftermath of the event, personal reactions to the event, and the reactions of others (Brown & Kulik, 1977). Over time, this set of canonical categories has expanded, incorporating additional details such as the timing of when individuals learned the news (Bohannon, 1988; Christianson, 1989; Neisser & Harsch, 1992), the presence of other individuals at the time of the news (Christianson, 1989; Kaya Kızılöz & Tekcan, 2013; Neisser & Harsch, 1992), the thoughts individuals had upon receiving the news (Christianson, 1989; Kaya Kızılöz & Tekcan, 2013; Weaver, 1993), and even contextual elements such as weather or clothing (Kaya Kızılöz & Tekcan, 2013; Kraha & Boals, 2014; Lanciano et al., 2018).

A methodological challenge in FBMs research is the assessment of their accuracy. Since FBMs are based on an individual's experience rather than on the public event itself, it is generally difficult to determine whether the remembered reception context accurately reflects the occurred events. In contrast to everyday autobiographical memories, which can sometimes be corroborated through objective evidence, FBMs typically concern personal events—the reception context of public news such as terrorist attacks, political assassinations, or disaster—making external validation problematic. Moreover, the majority of FBM research is retrospective in nature, which exacerbates this issue. Memories are more prone to distortion from post-event information, repeated rehearsal, and the impact of dominant societal and cultural narratives when they are evaluated a long time after the initial event. Even if people continue to believe that their FBMs are accurate and vivid, there is empirical evidence that these memories, like other autobiographical memories, are reconstructive and subject to change with time. Therefore, accuracy is not always correlated with high confidence (Talarico & Rubin, 2003; Neisser & Harsch, 1992). One major limitation of retrospective studies is the lack of control over the original encoding context. Researchers can neither verify whether participants truly experienced the event in the way they describe, nor whether their current memories are consistent with earlier undocumented recollections. This leads to a second methodological concern, such as the reliability on self-report measures, which may combine subjective confidence with objective accuracy. To address these challenges, researchers have advocated for more rigorous methodologies, including test–retest or longitudinal designs, and explicit and implicit measurement (Curci et al., 2015; Lanciano et al., 2013). Such approaches offer a more reliable way of assessing the accuracy and stability of FBMs and contribute to a more appropriate understanding of how personal and collective memory interact during experiences of major significance.

A substantial body of research on FBM assessment converges on three distinct parameters for FBM assessment (Bohannon & Symons, 1992): The quantity of details recalled (specificity), the degree of confidence individuals have regarding their memories (confidence), and the stability of these memories over time (consistency). The first parameter—FBM specificity—denotes the richness of details retrieved in the individual's memory, encompassing the number of canonical categories recalled in either spontaneous or probed reports (Bohannon, 1988; Bohannon & Symons, 1992; Brown & Kulik, 1977; Lanciano et al., 2013, 2019). Furthermore, some researchers have highlighted self-reported vividness as a phenomenological distinguishing characteristic of FBM richness (Demiray & Freund, 2015; Gandolphe & El Haj, 2017; Kvavilashvili et al., 2010; Niedźwieńska, 2003; Talarico & Rubin, 2003, 2007). FBMs exhibit a notable level of vividness, with a subjective sensation that the original event can be "re-experienced." As a result, memories are full of contextual details (e.g., time and space, other people present, location, or actions people took), perceptual/sensorial details (e.g., visual, tactile, auditory, and olfactory details), somatic/visceral/bodily details (e.g., heartbeat, perspiration, breath, and stomach), and affective details (e.g., initial thoughts people had or the emotions and the affective reactions people had).

FBM studies have assessed a strong sense of confidence often associated with vivid and lasting autobiographical memories. As a result, some scholars contend that this proxy, rather than consistency throughout time, is what primarily defines FBMs (Curci et al., 2015; Day & Ross, 2014; Neisser & Harsch, 1992; Talarico & Rubin, 2003). People may be quite confident that they can remember particular details from their initial experiences, however their memories may turn out to be false. In conclusion, a person's confidence of memory accuracy does not always coincide with its accuracy.

In the literature on autobiographical memory, consistency is widely used as the most reliable proxy for memory accuracy. Since empirical research often lacks direct access to the original autobiographical events, repeated recounting of the same event is believed to enhance recall accuracy. FBM consistency has been extensively validated through test–retest studies, in which participants recall their memories at multiple time points to assess both consistency and accuracy (Curci & Luminet, 2006). Recent research further supports FBM consistency as the strongest parameter associated with implicit measures obtained through the autobiographical Implicit Association Test (aIAT; Sartori et al., 2008; Curci et al., 2015; Lanciano et al., 2013).

4 FBMs for Covid-19 Events

The dynamic characteristics of the Covid-19 pandemic presented a unique opportunity to analyze FBMs outside the traditional paradigm of a single event. This unprecedented and continually changing emergency prompted a broader examination, incorporating several elements of the pandemic to understand how an enduring worldwide situation can affect related memories (Lanciano et al., 2024; Ma & Wan, 2022). In contrast to the sudden, clearly defined events that are typically linked to FBMs, such as terrorist attacks, natural disasters or the death of famous figures, the Covid-19 pandemic has been a protracted, complex crisis characterized by a series of emotionally charged episodes, including the first nationwide lockdowns, the declaration of alarming death tolls, pictures of overcrowded hospitals or the introduction of vaccines. Its extended temporal dimension challenges the classical conceptualization of FBMs as singular, idiosyncratic memories of brief, limited surprising events (Brown & Kulik, 1977). Instead, the pandemic probably gave rise to a series of snapshot moments of memories that were spaced out throughout time and connected to different emotional peaks. This changing memory perspective emphasizes how social dimensions and persistent emotional involvement help to reinforce and reshape FBMs over time (Finkenauer et al., 1998; Talarico et al., 2004). Indeed, the prolonged nature of the emergency not only impacts on the processes of memory encoding and retrieval, but also emphasizes the reconstructive nature of FBMs, which are influenced by individual experience and group meaning making (Hirst & Phelps, 2016). As a result, the Covid-19 offers an innovative frame for investigating FBMs in relation to highly defining specific moments within a very long and lasting-time window such as that of the pandemic. The selection of these target events is, quite evidently, influenced by the level of public concern or perceived threat associated with them. As such, emotional salience and importance appear to mediate which moments are encoded as *flashbulb target events* (e.g., the announcement of the first national lockdown or the confirmation of the initial case in a specific region), consistent with prior research indicating that emotional intensity and personal consequentiality are critical determinants in the formation of FBMs.

As support to this viewpoint, Jiménez Frei and Carlson (2021) coined the term *COVID flashbulb memories* in the school context to describe memories that teachers and students have of the decision to switch to online lessons in the spring of 2020. To our knowledge, few studies have investigated the FBM formation and maintenance for the Covid-19 emergency and pandemic (Herrmann et al., 2023; Lanciano et al., 2024; Ma & Wan, 2022; Xuan et al., 2024). Following the idea that personal relevance in the original event is a crucial antecedent for the FBM formation (*see Sect. 5.1*), Ma and Wan (2022) investigated the role of personal involvement on FBM for Covid-19 pandemic, by operationalizing it in terms of both spatiotemporal involvement (spatiotemporal distance) and empathic involvement (empathy level). The findings of a two-stage online survey conducted in Shanghai revealed that FBM specificity was impacted negatively by empathic involvement; that is, the more empathy one possessed, the lower one's FBM specificity would be. Additionally, the spatiotemporal engagement of individuals was found to significantly improve the FBM specificity and consistency.

Herrmann et al. (2023) investigated memories for the first Covid-19 lockdown in the Netherlands in a sample of patients with severe amnesia, due to Korsakoff's syndrome (KS). Twenty-two KS patients and twenty-four age-, education-, and gender-matched healthy controls took part in the study. Measures of emotional intensity, phenomenological features, autobiographical memory specificity, and semantic and episodic information of the initial lockdown in the Netherlands from March to May 2020 were assessed. Results showed that one-fourth of the KS patients reported specific details, although they remembered far less autobiographical information about the Covid-19 lockdown than healthy controls. KS patients exhibited levels of emotional intensity similar to those in the control group. Memories of lockdown-related episodic details were higher than memories of lockdown-related semantic details in both amnesia patients and healthy controls. The findings show that even people with severe amnesia can still encode, store, and recall knowledge about macro events, particularly emotionally charged ones.

Xuan et al. (2024) assessed FBMs and personal memories for the news of switching to online classes in Colby college due to Covid-19. This restrictive decision exceptionally presents a situation in which FBM and first-hand memories are related: All college members learned of the closure decision made by the college administration (i.e., FBM for this news) and were personally affected by the decision (i.e., first-hand memory for the closure events). The authors assessed respondents' opinions of the choice, its effects, and whether the news of the campus' closure led to FBMs. Through a two-wave longitudinal survey administered in

March and May 2020, participants answered about the news of the campus' closing and about a control memory (an event from the same week selected by participants). According to participant accounts, individuals produced FBMs, and these memories held up better over time than control memories. There was no difference in FBM confidence between memory types. Furthermore, the authors found a resoundingly positive reaction at first to the campus closure decision, which grew stronger over time. Finally, in the second wave of testing, participants' emotional responses shifted from negative emotions in the first wave to more neutral ones.

In a cross-national study (Canada, China, Denmark, Greece, Italy, Malaysia, New Zealand, Poland, Russia, Spain, and Turkey), Lanciano et al. (2024) asked their participants to remember the context in which they received news of the first Covid-19 case in their own country, i.e., the canonical categories: (1) day, month, and year they first heard of the first Covid-19 case, (2) where they were, (3) what they were doing, (4) the source of the news, and (5) other people present when the news was received. Additionally, the authors assessed pandemic-related emotional and cognitive factors, including both objective (such as governmental constraints) and subjective variables (i.e., Covid-19 subjective severity, pandemic impact expectedness, personal and global importance of pandemic, and rehearsal of pandemic news and events). The authors hypothesized that respondents would recall specific and detailed memories of the reception context, and that high levels of emotional involvement, along with significant Covid-19 severity, would correlate with increased FBM specificity. Furthermore, this study addressed the critical need to investigate FBMs for Covid-19 pandemic in countries showing different levels of concerns in the original events, operating under the hypothesis that shorter psychological and physical distances from a pivotal event (as measured by the incidence of Covid-19 cases, mortality rates, and governmental interventions) would have corresponded to heightened FBM specificity (Conway et al., 1994; Er, 2003; Finkenauer et al., 1998). The authors found that participants had partially specific memories of the location, ongoing activity, and the news source, and totally detailed memories about the day and other people present when they heard the news. China exhibited the highest FBM specificity, and all countries considered the Covid-19 situation to be of greatest importance on both a personal and global level.

Lanciano et al. (2024) also explored age and gender differences in FBM specificity and found that women and younger individuals tend to retain the learning context of the first Covid-19 case more vividly and with greater detail. These findings align well with previous empirical evidence for which younger adults generally show a greater propensity to develop and retain these memories compared to older individuals (Kopp et al., 2020). This disparity may be partly explained by differences in the cognitive and emotional mechanisms underpinning memory formation: Younger individuals tend to rely more on emotional connectedness to encode such memories, whereas older adults depend more heavily on rehearsal and are more likely to forget contextual details. Additionally, the widespread use of digital technologies, which enable individuals to document and share experiences instantly, has uniquely shaped the collective memory of Covid-19, distinguishing it from previous global events in both form and accessibility. About gender differences, Lanciano et al.' (2024) results correspond with evidence of a female advantage in episodic memory (Asperholm et al., 2019). In general, broader episodic memory patterns are reflected in gender variability in FBMs: males tend to do better in spatial memory, whereas women generally perform better in verbal and sensory memory tasks (Asperholm et al., 2019). Moreover, studies indicate that after traumatic experiences like the Covid-19 epidemic, women are more prone than males to have intrusive memories (Kendler et al., 2001). Furthermore, men and women differ not only in how they create and narrate their memories, but also in the content of their recollections (Niedźwieńska, 2003).

Very recently, Aizpurua Sanz et al. (2025) examined FBMs related to the declaration of the Covid-19 alarm state in Spain, focusing on age-related differences in memory specificity and confidence. Their findings revealed that the event fulfilled the core criteria of FBMs, with participants recalling highly specific and confident memories. Regarding age differences, younger individuals provided more concrete and detailed recollections than middle-aged and older adults, although there were no significant differences in the confidence levels across age groups.

5 FBMs Formation and Maintenance

Why should the Covid-19 pandemic lead people to produce an FBM? What are the conditions and characteristics of an event that cause this special type of memory? A large body of studies has investigated the circumstances that lead to the formation of FBMs and the factors that impact their maintenance over time (Brown & Kulik, 1977; Conway et al., 1994; Curci & Luminet, 2009; Er, 2003; Finkenauer et al., 1998; Luminet & Curci, 2009). The notion of FBM as a special type of memory, distinct from ordinary autobiographical recollections, has sparked an enduring debate among supporters and critics of this perspective (see Curci & Conway, 2013). Since Brown and Kulik's pioneering article (1977), several models of FBMs (Brown & Kulik, 1977; Conway et al., 1994; Finkenauer et al., 1998; Er, 2003) have emphasized the theoretical link between memory formation and cognitive, emotional, and social determinants. The discussion centers on whether a special encoding mechanism is indeed activated for the FBM formation and maintenance, or whether FBMs can be explained through ordinary autobiographical memory processes. This debate still remains unresolved.

According to proponents of the special-mechanism hypothesis, FBMs arise due to unique encoding triggered by emotional intensity, surprise, and consequentiality. These factors result in vivid, long-lasting memories, often described as having a "live quality" (Brown & Kulik, 1977), alike to a camera-like mechanism that captures a moment with rich sensory and perceptual detail. Critics of the special-mechanism view, such as Neisser (1982) and Rubin and Kozin (1984), questioned the accuracy of FBMs and challenged the idea that they are categorically distinct from other autobiographical memories. They argue that, although FBMs are experienced with high confidence, they are still reconstructive and prone to distortion

(Neisser & Harsch, 1992; Wright & Gaskell, 1995). The ordinary-memory perspective proposes that FBMs result from standard autobiographical processes, heavily influenced by post-event factors such as social sharing, narrative construction, and rehearsal. This contradiction suggests that FBMs may best be understood through an integrated approach that bridges the two perspectives, viewing them as dynamic phenomena influenced both by encoding emotional conditions and by post-encoding social processes (Curci & Luminet, 2006; Finkenauer et al., 1998; Lanciano et al., 2010).

5.1 The Role of Emotional Encoding Factors in FBMs

The emotional impact of the event, the degree of surprise and novelty, as well as the personal and social relevance of an event are some encoding factors leading to a FBM (Brown & Kulik, 1977; Conway et al., 1994). This hypothesis surrounding a special encoding proposes that FBMs serve as relatively accurate representations of the initial event. The special encoding hypothesis is based on the *"Now Print!"* mechanism (Livingston, 1967), a neurological process assumed to have the evolutionary purpose to enable our ancestors to recognize harmful occurrences based on their environmental correlates. When an event happens unexpectedly and has biologically important effects on a person's life, the *Now Print!* mechanism is triggered so that the person is ready to identify such situations in the future. People are able to recall not just the actual event, but also unimportant and incidental details, such the circumstances surrounding their initial news exposure.

Conway et al. (1994) supported Brown and Kulik's (1977) idea of FBMs as a special class of memory. The authors examined a large sample of UK and non-UK residents in order to explore the unexpected resignation of the British Prime Minister Thatcher in 1990. They believed that the prevalence of extremely detailed and consistent memory reports demonstrated that FBM was a special type of memory. The authors also agreed with Brown and Kulik that an event needs to become significant enough to set off an FBM. They state that *"the item of news has to be judged as having consequences for self that are more significant than the consequences typically assigned to most items of news"* (Conway et al., 1994, p. 339). Pillemer (1984) also believed in the specific encoding hypothesis, underpinning the crucial importance of emotions and surprise in remembering the murder attempt on Ronald Reagan. Bohannon (1988) asserted that both emotional intensity and rehearsal are crucial factors in the development and upkeep of FBM. Both must work together for FBM to remain stable over time: The emotional impact of an event makes it possible for individuals to think and share it, making memory of the event more long-lasting.

The importance—consequentiality has also been identified as a crucial factor related to memory for the reception context, causing subsequent emotional response to the event (Brown & Kulik, 1977; Christianson, 1989; Conway et al., 1994; Finkenauer et al., 1998; Rubin & Kozin, 1984). The significance and consequences of events can be interpreted in various manners. Some scholars have examined this dimension through the lens of social group affiliation, which pertains to specific levels of concern regarding the initial event (e.g., social or national identity, political issues, religious connections; Curci et al., 2015; Curci & Luminet, 2009; Kvavilashvili et al., 2009; Tinti et al., 2009). In contrast, other researchers have framed the evaluation of importance and consequentiality as reliant on emotional engagement, characterized by direct experience of the event (Er, 2003; Neisser et al., 1996) or the psychological proximity to the event (Tekcan et al., 2003). Studies analyzing FBMs and Event Memories (EMs; the recollection of the target event details) across different social groups revealed that individuals store and sustain their memories of an event along with its reception context based on their level of emotional involvement and interest (Curci et al., 2001; Curci & Luminet, 2009).

In their study, Tinti et al. (2009) investigated the extent to which the FBMs for the death of Pope John Paul II differ according to the persons' evaluation of the event's importance and consequentiality. FBMs were assessed in individuals with varying levels of importance—consequentiality based on two variables: religious participation and nationality (Polish, Italian, Swiss). The findings showed that importance—consequentiality is a key factor in FBM and is impacted by social antecedent and personal traits. Furthermore, while surprise does not appear to be a crucial factor in FBM, memory consistency appears to be impacted both directly by emotional intensity and indirectly through rehearsal.

Furthermore, Curci et al. (2015) assessed FBMs and EMs for the news of Pope Benedict XVI's resignation in three samples of Catholic churchgoers, non-churchgoers, and Evangelical Italian participants using explicit (FBM Specificity, Confidence, Consistency) and implicit (autobiographical Implicit Association Test, aIAT; Sartori et al., 2008) measures. The results confirmed the substantial correlation between accurate EM details and FBM. The aIAT impact was higher for FBMs than for EM details among Catholic respondents, who were the group most impacted by the news.

To assess the FBM Consistency over time, Cordonnier and Luminet (2021) conducted a test–retest study on the 2016 Brussels bombings with the following objectives: (1) collect baseline data on the day of the event and retest data using an in-depth interview approach; and (2) evaluate social identification at different levels (local-Brussels, national-Belgium, and supranational-Europe). The authors evaluated pride and tie to the nation in addition to the conventional operationalization of social identity in terms of participants' citizenship (e.g., Curci & Luminet, 2006; Curci et al., 2001; Talarico et al., 2019). In the particular context of the Brussels bombings, social identification with Brussels and Europe correlated with FBM indices, while social identification with Belgium did not.

5.2 The Role of Social Post-Encoding Factors in FBMs

Besides the encoding factors approach to explain FBMs, an alternative viewpoint emphasizes the impact of post-encoding social factors (Wright & Gaskell, 1995). According to research on collective memory, shared events take place in social settings might contribute to FBMs research (Pennebaker et al., 1997). This reconstructive hypothesis supports the role of rehearsal mechanisms such as rumination, social interaction, seeking news, prior knowledge of the event, and attitude toward the event (Christianson, 1989; McCloskey et al., 1988; Neisser & Harsch, 1992; Pezdek, 2003; Talarico & Rubin, 2003). In this vein, the FBM formation aligns with a reconstructive framework, conforming to narrative schemas or scripts (Neisser, 1982; Neisser & Harsch, 1992). McCloskey et al. (1988) came to the conclusion that there is no need to postulate a special mechanism, rather, FBM should be considered a byproduct of "normal" autobiographical memory processes.

According to this reconstructive approach, the rehearsal of the original event plays a crucial role in FBMs formation and maintenance. Rehearsal encompasses different processes such as searching and following event-related news through the media, sharing the event with others, and thinking of the original event (Bohannon & Symons, 1992; Christianson, 1989; Er, 2003; Finkenauer et al., 1998; McCloskey et al., 1988; Neisser, 1982). Although the initial memory trace remains constant and detailed throughout time, memory reports change, depending on the subsequent exposure to overt (social sharing and mass media) and covert (mental rumination) rehearsal processes (Brown & Kulik, 1977). The significant impact of the rehearsal process on the FBM vividness was also recognized by Cheriet et al. (2023) in a recent study exploring FBMs, collective identity, future thinking, and shared representations for the Capitol Riots event, which happened in Washington DC, on 6 January 2021. Among other results, findings revealed that Americans reported more FBMs than Belgians and underlined the importance of rehearsal through media and communication in FBM formation.

In analogy, the Covid-19 crisis starkly illustrated the profound consequences associated with information-seeking behaviors, leading to phenomena such as information overload and infodemics (Rathore & Farooq, 2020; Luo et al., 2022). Any retrospective evaluations about the suitability of the measures taken and the efforts made to enhance future pandemic preparedness depends on the accurate recall of data and information people have. These memories and retrospective evaluations are greatly biased by public and media discourses (Sprengholz et al., 2023).

5.3 The FBM Emotional-Integrative Model

Finkenauer et al. (1998) attempted to reconcile the debate about special vs. not special mechanism by introducing their emotional-integrative model. This model suggests that the interplay between emotion and FBM can occur both directly—through the evaluation of novelty and surprise concerning the original event—and indirectly—via rehearsal processes (Curci & Luminet, 2006; Lanciano et al., 2010). According to the authors: (1) a reaction of surprise results from the appraisal of the original event in terms of novelty; (2) an emotional feeling state results from the surprise and from the appraisal of the original events in terms of importance/consequentiality; (3) the emotional feeling state may affect FBM directly or indirectly by causing rehearsal, which in turn affects the memory for the original event. In summary, the emotional-integrative model postulated that there might be a direct (by the evaluation of novelty and surprise) or indirect (via rehearsal) path between the emotional feeling state and FBM.

The emotional-integrative model has been confirmed as a hopeful framework to account FBMs. Curci and Luminet (2009) applied the model to test the FBMs for the news of the death of President Mitterrand, across two French and Belgian groups. Results showed that FBMs are predicted by their emotional impact and subsequent rehearsal, while surprise did not work. Furthermore, Luminet and Curci (2009) examined four theoretical models to explain the FBM formation (Brown & Kulik, 1977; Conway et al., 1994; Er, 2003; Finkenauer et al., 1998) in two US and non-US groups; on average 21 and 524 days after the 9/11 attacks. The findings revealed that the "direct path" (from novelty/surprise to FBMs) of the Finkenauer and colleagues' model was only significant for the US group, whereas the "indirect path" (from emotional states, rehearsal, and event–memory to FBMs) was only significant for the non-US group. This implies that the direct path cannot be initiated without, first, activation of social identity: Only when an event is relevant to a social community, it can create lasting FBMs, as in the case of the social identity of the US people, strongly triggered by the 9/11 attacks.

6 Conclusive Remarks: Covid-19 as Both Personal and Collective Memory

Across the realm of autobiographical remembering, the unusual and unexpected global outbreak of Covid-19 provided us with a rare chance to investigate the emotional effects on personal and collective memories (Topçu & Hirst, 2024). The Covid-19 pandemic is obviously not easily characterized as other well-known public events, such as 9/11 (Hirst et al., 2009, 2015) or the fall of the Berlin Wall (Bohn & Berntsen, 2007). One significant difference is the pandemic's longer duration, which means that every local outbreak report or lockdown announcement could have potentially led to the creation of a FBM or act as a memory of past events. Few studies have examined long-term and ongoing events, like the Covid-19 pandemic, in contrast to all FBMs studies, which have focused on events that happened in the past, had distinct beginnings, and ends (e.g., Curci et al., 2015, 2020; Curci & Lanciano, 2009; Lanciano et al., 2013; Lanciano & Curci, 2012; Tinti et al., 2009).

In this vein, FBMs for Covid-19 pandemic represent a perfect middle ground between a flashbulb episodic memory in the strict sense and a collective memory. The unprecedented uniqueness of the pandemic has made many of the Covid-19 events feel like "*first-time experiences*." As a result, this may have led to the formation of a photographic memory of the context in which individuals first learned about key news, such as the initial Covid case, the first lockdown, or the announcement of school closures. However, at the same time, the prolonged and continuous nature of the Covid-19 events, their repeated and continuous occurrence, their sharing around the world, make these memories as *life periods* remembered at both individual and collective levels (Luo et al., 2022; Mustafa et al., 2021). In this regard, Hirst (2025) speaks of *flashbulb memories as sites for collective memories* (p. 85). After 10 years, 99.5% of the American sample still has a FBM for the 9/11 attacks (Hirst et al., 2015), and similarly, 98.8% of sample still has a FBM for Kennedy's assassination (Brown & Kulik, 1977). These two findings imply that FBMs may be used as a sign of community affiliation. In other words, FBMs have the ability to anchor both individual and community identity. Neisser (1982) asserts that self-relevant details are well retained in memory because an event serves as a "benchmark," allowing people to declare, *"I was there"* at pivotal moments in history, and marks the intersection of personal and collective history (Neisser, 1982). When a FBM is formed, the memory of the event becomes more distinctively community-based, more crucial in confirming one's community identity, and more memorable. Naturally, not every long-lasting collective memory has caused FBMs in the general population as it developed. However, FBMs are ways of bringing an event that is shared by many people into a community's collective memory (Hirst, 2025).

According to the Transition Theory (TT), the impact that public events have on everyday routines determines how memorable such events are as well (Brown, 2016). TT supposes that the structure of experience is reflected in the content and organization of autobiographical memory. The *living-in-history* (Brown et al., 2009) effect has been found, showing that individuals who have survived long-lasting wars or natural disasters are more likely to use public events as an organizational framework for their personal life stories than those who have merely observed significant events. According to this perspective, transitions are events or sets of events that lead to the start and finish of lifetime periods, and provide the circumstances necessary for the emergence of a new set of event components. In keeping with this, Covid-19 would be the biggest global collective transition that the world has ever undergone (Brown, 2021). All people around the world have been somewhat altered from their pre-pandemic lifestyles as a result of being under quarantine, lockdown, or self-isolation. As a result, people mention and will continue to mention the pandemic as a temporal landmark when dating events (Ekinci & Brown, 2024), a sort of *b.C.* (before Covid-19) and *a.C.* (after Covid-19). The high memorability of the Covid-19 pandemic at both individual and collective levels would be important evidence that people will build their lives post-pandemic based on not only positive life events—as the cultural life script theory suggests (Berntsen & Rubin, 2004; Rubin & Berntsen, 2003)—but also around negative ones.

The studies on FBMs for Covid-19 events, although still limited, allow us to get a more comprehensive view of what (memory content) and how (memory organization) people remember emotionally charged events such as Covid-19, paying great attention to the emotional, cognitive, and social worldwide effects. Past remembering may help people to connect past and future, to promote and enhance coping and resilience strategies, and to regulate emotions (Öner & Gülgöz, 2018). According to Bluck et al. (2005), memory serves a number of psychological functions in a person's life, including: (1) directing behavior in the future by learning from the past (directive function), (2) maintaining a cohesive sense of self (self-function), and (3) preserving social connections (social function). Based on Bluck theory, Adıgüzel et al. (2024) proposed other autobiographical memory functions, such as the emotion regulation function (Öner & Gülgöz, 2018) and the predictive function (i.e., predicting the future by recollecting the past; Ay & Gülgöz, 2024). While the directive function focuses more on controlling the future based on lessons gained from the past, the predictive function helps eliminate ambiguity and uncertainty by using prior experiences, as in the Covid-19 case (Ay & Gülgöz, 2024). The emotion regulation function was most frequent in forecasting post-pandemic events, indicating that pandemic future simulations may be used to relieve the emotional stress that is currently being experienced (Adıgüzel et al., 2024).

In conclusion, future studies should continue to investigate how the Covid-19 pandemic acts as a landmark to organize the content and structure of both personal and collective memories. Since memories and forecasting thinking are linked to psychological adjustment to personal events (Banks & Salmon, 2013; Conway & Pleydell-Pearce, 2000), how individuals remember their experiences and envision their future during the Covid-19 pandemic may be strictly linked to their health. Further research needed to be done to determine whether people's psychological well-being and ability to cope with challenging circumstances are related to the semantic and episodic elements of remembering pandemic period, like Covid-19.

References

Adams, T., & Kopelman, S. (2021). Remembering COVID-19: Memory, crisis, and social media. *Media, Culture & Society, 44*(2), 266–285. https://doi.org/10.1177/01634437211048377

Adıgüzel, Z., Ay, D., Bilgin, E., Coşkuner, S. B., Ergen, İ., & Gülgöz, S. (2024). The functions of remembering the past and imagining the future during the COVID-19 pandemic. *Applied Cognitive Psychology, 38*(1), e4181. https://doi.org/10.1002/acp.4181

Aizpurua Sanz, A., Migueles Seco, M., & Unibaso-Markaida, I. (2025). Flashbulb memories for the declaration of the COVID-19 alarm state: Age-related differences. *Psychological Research, 89*(109), 1–11. https://doi.org/10.1007/s00426-025-02140-1

Asperholm, M., Högman, N., Rafi, J., & Herlitz, A. (2019). What did you do yesterday? A meta-analysis of sex differences in episodic memory. *Psychological Bulletin, 145*(8), 785–821. https://doi.org/10.1037/bul0000197

Ay, D., & Gülgöz, S. (2024). Characteristics and functions of predictive and directive memories and forecasts. *Applied Cognitive Psychology, 38*(1), e4174. https://doi.org/10.1002/acp.4174

Banks, M. V., & Salmon, K. (2013). Reasoning about the self in positive and negative ways: Relationship to psychological functioning in young adulthood. *Memory, 21*(1), 10–26. https://doi.org/10.1080/09658211.2012.707213

Berntsen, D., & Rubin, D. C. (2004). Cultural life scripts structure recall from autobiographical memory. *Memory & Cognition, 32*(3), 427–442. https://doi.org/10.3758/BF03195836

Bluck, S., Alea, N., Habermas, T., & Rubin, D. C. (2005). A tale of three functions: The self-reported uses of autobiographical memory. *Social Cognition, 23*(1), 91–117. https://doi.org/10.1521/soco.23.1.91.59198

Bohannon, J. N. (1988). Flashbulb memories for the space shuttle disaster: A tale of two theories. *Cognition, 29*(2), 179–196. https://doi.org/10.1016/0010-0277(88)90036-4

Bohannon, J. N., & Symons, V. L. (1992). Flashbulb memories: Confidence, consistency, and quantity. In E. Winograd & U. Neisser (Eds.), *Affect and accuracy in recall* (1st ed., pp. 65–92). Cambridge University Press. https://doi.org/10.1017/CBO9780511664069.005

Bohn, A., & Berntsen, D. (2007). Pleasantness bias in flashbulb memories: Positive and negative flashbulb memories of the fall of the Berlin Wall among East and West Germans. *Memory & Cognition, 35*(3), 565–577. https://doi.org/10.3758/BF03193295

Brown, N. R. (2016). Transition theory: A minimalist perspective on the organization of autobiographical memory. *Journal of Applied Research in Memory and Cognition, 5*(2), 128–134. https://doi.org/10.1016/j.jarmac.2016.03.005

Brown, N. R. (2021). The possible effects of the COVID-19 pandemic on the contents and organization of autobiographical memory: A transition-theory perspective. *Cognition, 212*, e104694. https://doi.org/10.1016/j.cognition.2021.104694

Brown, N. R., Lee, P. J., Krslak, M., Conrad, F. G., Hansen, T. G. B., Havelka, J., & Reddon, J. R. (2009). Living in history: How war, terrorism, and natural disaster affect the organization of autobiographical memory. *Psychological Science, 20*(4), 399–405. https://doi.org/10.1111/j.1467-9280.2009.02307.x

Brown, R., & Kulik, J. (1977). Flashbulb memories. *Cognition, 5*(1), 73–99. https://doi.org/10.1016/0010-0277(77)90018-X

Castillo, J., Fan, H., Karaman, O. T., Shu, J., Stussi, Y., Kredlow, M. A., et al. (2024). Overestimating the intensity of negative feelings in autobiographical memory: Evidence from the 9/11 attack and Covid-19 pandemic. *Cognition and Emotion, 38*(7), 1048–1063. https://doi.org/10.1080/02699931.2024.2346757

Chamaa, F., Bahmad, H. F., Darwish, B., Kobeissi, J. M., Hoballah, M., Nassif, S. B., Ghandour, Y., Saliba, J.-P., Lawand, N., & Abou-Kheir, W. (2021). PTSD in the COVID-19 era. *Current Neuropharmacology, 19*(12), 2164–2179. https://doi.org/10.2174/1570159X19666210113152954

Cheriet, N., Topçu, M., Hirst, W., Bastin, C., & Folville, A. (2023). A day that America will remember: Flashbulb memory, collective memory, and future thinking for the capitol riots. *Memory, 31*(5), 715–731. https://doi.org/10.1080/09658211.2023.2190570

Christianson, S.-Á. (1989). Flashbulb memories: Special, but not so special. *Memory & Cognition, 17*(4), 435–443. https://doi.org/10.3758/BF03202615

Cole, S. N., Markostamou, I., Watson, L. A., Barzykowski, K., Ergen, İ., Taylor, A., & Öner, S. (2023). Spontaneous past and future thinking about the COVID-19 pandemic across 14 countries: Effects of individual and country-level COVID-19 impact indicators. *Journal of Applied Research in Memory and Cognition, 12*(4), 502–512. https://doi.org/10.1037/mac0000071

Conway, M. A., & Pleydell-Pearce, C. W. (2000). The construction of autobiographical memories in the self-memory system. *Psychological review, 107*(2), 261–288. https://doi.org/10.1037/0033-295X.107.2.261

Conway, M. A., Anderson, S. J., Larsen, S. F., Donnelly, C. M., McDaniel, M. A., McClelland, A. G. R., Rawles, R. E., & Logie, R. H. (1994). The formation of flashbulb memories. *Memory & Cognition, 22*(3), 326–343. https://doi.org/10.3758/BF03200860

Cordonnier, A., & Luminet, O. (2021). Consistency and social identification: A test-retest study of flashbulb memories collected on the day of the 2016 Brussels bombings. *Memory, 29*(3), 305–318. https://doi.org/10.1080/09658211.2021.1891253

Curci, A., & Conway, M. A. (2013). Playing the flashbulb memory game: A comment on Cubelli and Della Sala. *Cortex, 49*(1), *352–355*. https://doi.org/10.1016/j.cortex.2012.05.004

Curci, A., & Lanciano, T. (2009). Features of autobiographical memory: Theoretical and empirical issues in the measurement of flashbulb memory. *Journal of General Psychology, 136*(2), 129–152. https://doi.org/10.3200/GENP.136.2.129-152

Curci, A., & Luminet, O. (2006). Follow-up of a cross-national comparison on flashbulb and event memory for the September 11th attacks. *Memory, 14*(3), 329–344. https://doi.org/10.1080/09658210500340816

Curci, A., & Luminet, O. (2009). Flashbulb memories for expected events: A test of the emotional-integrative model. *Applied Cognitive Psychology, 23*(1), 98–114. https://doi.org/10.1002/acp.1444

Curci, A., Lanciano, T., Curtotti, D., & Sartori, G. (2020). Lessons for the courtroom from the study of flashbulb memory: An integrative review. *Memory, 28*(3), 441–449. https://doi.org/10.1080/09658211.2020.1727522

Curci, A., Lanciano, T., Maddalena, C., Mastandrea, S., & Sartori, G. (2015). Flashbulb memories of the Pope's resignation: Explicit and implicit measures across differing religious groups. *Memory, 23*(4), 529–544. https://doi.org/10.1080/09658211.2014.908923

Curci, A., Luminet, O., Finkenauer, C., & Gisle, L. (2001). Flashbulb memories in social groups: A comparative test–retest study of the memory of French President Mitterrand's death in a French and a Belgian group. *Memory, 9*(2), 81–101. https://doi.org/10.1080/09658210042000120

Day, M. V., & Ross, M. (2014). Predicting confidence in flashbulb memories. *Memory, 22*(3), 232–242. https://doi.org/10.1080/09658211.2013.778290

Demiray, B., & Freund, A. M. (2015). Michael Jackson, Bin Laden and I: Functions of positive and negative, public and private flashbulb memories. *Memory, 23*(4), 487–506. https://doi.org/10.1080/09658211.2014.907428

Er, N. (2003). A new flashbulb memory model applied to the Marmara earthquake. *Applied Cognitive Psychology, 17*(5), 503–517. https://doi.org/10.1002/acp.870

Ekinci, Ö., & Brown, N. R. (2024). The COVID-19 pandemic as autobiographical period: Evidence from an event dating study. *Memory, 32*(2), 283–291. https://doi.org/10.1080/09658211.2024.2310562

Finkenauer, C., Luminet, O., Gisle, L., El-Ahmadi, A., Van Der Linden, M., & Philippot, P. (1998). Flashbulb memories and the underlying mechanisms of their formation: Toward an emotional-integrative model. *Memory & Cognition, 26*(3), 516–531. https://doi.org/10.3758/BF03201160

Gandolphe, M.-C., & El Haj, M. (2017). Flashbulb memories of the Paris attacks. *Scandinavian Journal of Psychology, 58*(3), 199–204. https://doi.org/10.1111/sjop.12364

Habermas, T. (2012). Identity, emotion, and the social matrix of autobiographical memory: A psychoanalytic narrative view. In D. C. Rubin & D. Berntsen (Eds.), *Understanding autobiographical memory: Theories and approaches* (pp. 33–53). Cambridge University Press. https://doi.org/10.1017/CBO9781139021937.005

Herrmann, D., Oudman, E., & Postma, A. (2023). The era of our lives: The memory of Korsakoff patients for the first COVID-19 pandemic lockdown in the Netherlands. *Consciousness and Cognition, 107*, 103454. https://doi.org/10.1016/j.concog.2022.103454

Hirst, W. (2025). Flashbulb memories as sites for collective memorie. In A. Erll, S. Knittel, & J. Wüstenberg (Eds.), *Dynamics, mediation, mobilization: Doing memory studies with Ann Rigney* (pp. 85–90). De Gruyter. https://doi.org/10.1515/9783111439273-012

Hirst, W., & Phelps, E. A. (2016). Flashbulb memories. *Current Directions in Psychological Science, 25*(1), 36–41. https://doi.org/10.1177/0963721415622487

Hirst, W., Phelps, E. A., Buckner, R. L., Budson, A. E., Cuc, A., Gabrieli, J. D., et al. (2009). Long-term memory for the terrorist attack of September 11: Flashbulb memories, event memories, and the factors that influence their retention. *Journal of Experimental Psychology: General, 138*(2), 161–176. https://doi.org/10.1037/a0015527

Hirst, W., Phelps, E. A., Meksin, R., Vaidya, C. J., Johnson, M. K., Mitchell, K. J., et al. (2015). A ten-year follow-up of a study of memory for the attack of September 11, 2001: Flashbulb memories and memories for flashbulb events. *Journal of Experimental Psychology: General, 144*(3), 604–623. https://doi.org/10.1037/xge0000055

Hjuler, T. F., Lee, D., & Ghetti, S. (2025). Remembering history: Autobiographical memory for the COVID-19 pandemic lockdowns, psychological adjustment, and their relation over time. *Child Development, 96*(1), 55–70. https://doi.org/10.1111/cdev.14131

Jacobson, N. C., Lekkas, D., Price, G., Heinz, M. V., Song, M., O'Malley, A. J., & Barr, P. J. (2020). Flattening the mental health curve: COVID-19 stay-at-home orders are associated with alterations in mental health search behavior in the United States. *JMIR Mental Health, 7*(6), e19347. https://doi.org/10.2196/19347

Jiménez Frei, C., & Carlson, S. (2021). Surviving, learning, and striving in the times of pandemic: Teaching with a journal of the plague year: An archive of COVID-19 (JOTPY). *Collections, 17*(3), 255–266. https://doi.org/10.1177/1550190620981031

Kaya Kızılöz, B., & Tekcan, A. I. (2013). Canonical categories in flashbulb memories. *Applied Cognitive Psychology, 27*(3), 352–359. https://doi.org/10.1002/acp.2913

Kendler, K. S., Thornton, L. M., & Prescott, C. A. (2001). Gender differences in the rates of exposure to stressful life events and sensitivity to their depressogenic effects. *American Journal of Psychiatry, 158*(4), 587–593. https://doi.org/10.1176/appi.ajp.158.4.587

Kopp, S. J., Sockol, L. E., & Multhaup, K. S. (2020). Age-related differences in flashbulb memories: A meta-analysis. *Psychology and aging, 35*(4), 459–472. https://doi.org/10.1037/pag0000467

Kraha, A., & Boals, A. (2014). Why so negative? Positive flashbulb memories for a personal event. *Memory, 22*(4), 442–449. https://doi.org/10.1080/09658211.2013.798121

Kvavilashvili, L., Mirani, J., Schlagman, S., Erskine, J. A. K., & Kornbrot, D. E. (2010). Effects of age on phenomenology and consistency of flashbulb memories of September 11 and a staged control event. *Psychology and Aging, 25*(2), 391–404. https://doi.org/10.1037/a0017532

Kvavilashvili, L., Mirani, J., Schlagman, S., Foley, K., & Kornbrot, D. E. (2009). Consistency of flashbulb memories of September 11 over long delays: Implications for consolidation and wrong time slice hypotheses. *Journal of Memory and Language, 61*(4), 556–572. https://doi.org/10.1016/j.jml.2009.07.004

Lanciano, T., & Curci, A. (2012). Type or dimension? A taxometric investigation of flashbulb memories. *Memory, 20*(2), 177–188. https://doi.org/10.1080/09658211.2011.651088

Lanciano, T., Alfeo, F., Curci, A., Marin, C., D'Uggento, A. M., Decarolis, D., et al. (2024). The flashbulb-like nature of memory for the first COVID-19 case and the impact of the emergency. A cross-national survey. *Memory, 32*(2), 264–282. https://doi.org/10.1080/09658211.2024.2310554

Lanciano, T., Curci, A., & Basile, P. (2019). Do psychopathic traits impair autobiographical memory for emotional life experiences? *Memory, 27*(5), 660–672. https://doi.org/10.1080/09658211.2018.1551493

Lanciano, T., Curci, A., Mastandrea, S., & Sartori, G. (2013). Do automatic mental associations detect a flashbulb memory? *Memory, 21*(4), 482–493. https://doi.org/10.1080/09658211.2012.740050

Lanciano, T., Curci, A., Matera, G., & Sartori, G. (2018). Measuring the flashbulb-like nature of memories for private events: The flashbulb memory checklist. *Memory, 26*(8), 1053–1064. https://doi.org/10.1080/09658211.2018.1428348

Lanciano, T., Curci, A., & Semin, G. R. (2010). The emotional and reconstructive determinants of emotional memories: An experimental approach to flashbulb memory investigation. *Memory, 18*(5), 473–485. https://doi.org/10.1080/09658211003762076

Lanciano, T., Graziano, G., Curci, A., Costadura, S., & Monaco, A. (2020). Risk perceptions and psychological effects during the Italian COVID-19 emergency. *Frontiers in Psychology, 11*, e580053. https://doi.org/10.3389/fpsyg.2020.580053

Liashenko, A., Dizaji, A. S., Melloni, L., & Schwiedrzik, C. M. (2020). Memory guidance of value-based decision making at an abstract level of representation. *Scientific Reports, 10*(1), 21496. https://doi.org/10.1038/s41598-020-78460-6

Liu, C. H., Zhang, E., Wong, G. T. F., Hyun, S., & Hahm, H. C. (2020). Factors associated with depression, anxiety, and PTSD symptomatology during the COVID-19 pandemic: Clinical implications for U. S. young adult mental health. *Psychiatry Research, 290*, e113172. https://doi.org/10.1016/j.psychres.2020.113172

Livingston, R. (1967). Reinforcement. In G. Quarton, T. Melenchunk, & F. Schmitt (Eds.), *The neurosciences: A study program* (pp. 568–576). Rockefeller University Press.

Luminet, O., & Curci, A. (2009). The 9/11 attacks inside and outside the US: Testing four models of flashbulb memory formation across groups and the specific effects of social identity. *Memory, 17*(7), 742–759. https://doi.org/10.1080/09658210903081827

Luo, R., Feng, W., & Xu, Y. (2022). Collective memory and social imaginaries of the epidemic situation in COVID-19—based on the qualitative research of college students in Wuhan, China. *Frontiers in Psychology, 13*, 998121. https://doi.org/10.3389/fpsyg.2022.998121

Lutz, C. S., Huynh, M. P., Schroeder, M., Anyatonwu, S., Dahlgren, F. S., Danyluk, G., Fernandez, D., Greene, S. K., Kipshidze, N., Liu, L., Mgbere, O., McHugh, L. A., Myers, J. F., Siniscalchi, A., Sullivan, A. D., West, N., Johansson, M. A., & Biggerstaff, M. (2019). Applying infectious disease forecasting to public health: A path forward using influenza forecasting examples. *BMC Public Health, 19*(1), 1659. https://doi.org/10.1186/s12889-019-7966-8

Ma, X., & Wan, J. (2022). Exploring the role of digital media dependency on the relationship between personal involvement and flashbulb memory during the pandemic: Empirical evidence from Mainland China. *Frontiers in Psychology, 13*, 985287. https://doi.org/10.3389/fpsyg.2022.985287

McCloskey, M., Wible, C. G., & Cohen, N. J. (1988). Is there a special flashbulb-memory mechanism? *Journal of Experimental Psychology: General, 117*(2), 171–181. https://doi.org/10.1037/0096-3445.117.2.171

Mustafa, H., Mukhiar, S. N. S., Jamaludin, S. S. S., & Jais, N. M. (2021). Covid-19 and collective memory among Malaysians: Does generation matter? *Pertanika Journal of Social Sciences & Humanities, 29*(4), 2371–2387. https://doi.org/10.47836/pjssh.29.4.15

Neisser, U. (1982). Snapshots or benchmarks? In U. Neisser & I. E. Hyman (Eds.), *Memory observed: Remembering in natural contexts* (pp. 68–74). Worth Publishers.

Neisser, U., & Harsch, N. (1992). Phantom flashbulbs: False recollections of hearing the news about challenger. In E. Winograd & U. Neisser (Eds.), *Affect and accuracy in recall* (1st ed., pp. 9–31). Cambridge University Press. https://doi.org/10.1017/CBO9780511664069.003

Neisser, U., Boodoo, G., Bouchard, T. J., Boykin, A. W., Brody, N., Ceci, S. J., Halpern, D. F., Loehlin, J. C., Perloff, R., Sternberg, R. J., & Urbina, S. (1996). Intelligence: Knowns and unknowns. *American Psychologist, 51*(2), 77–101. https://doi.org/10.1037/0003-066X.51.2.77

Niedźwieńska, A. (2003). Gender differences in vivid memories. *Sex Roles, 49*(7), 321–331. https://doi.org/10.1023/A:1025156019547

Öner, S., & Gülgöz, S. (2018). Autobiographical remembering regulates emotions: A functional perspective. *Memory, 26*(1), 15–28. https://doi.org/10.1080/09658211.2017.1316510

Öner, S., Watson, L. A., Adıgüzel, Z., Ergen, İ., Bilgin, E., Curci, A., Cole, S., De La Mata, M. L., Janssen, S. M. J., Lanciano, T., Markostamou, I., Nourkova, V., Santamaría, A., Taylor, A., Barzykowski, K., Bascón, M., Bermeitinger, C., Cubero-Pérez, R., & Dessenberger, S.,…Uner, O. (2022). Collective remembering and future forecasting during the COVID-19 pandemic: How the impact of COVID-19 affected the themes and phenomenology of global and national memories across 15 countries. *Memory & Cognition, 51*(3), 729–751. https://doi.org/10.3758/s13421-022-01329-8

Pennebaker, J. W., Paez, D., & Rimé, B. (Eds.). (1997). *Collective memories of political events: Social psychological perspectives*. Erlbaum.

Pezdek, K. (2003). Event memory and autobiographical memory for the events of September 11, 2001. *Applied Cognitive Psychology, 17*(9), 1033–1045. https://doi.org/10.1002/acp.984

Pillemer, D. B. (1984). Flashbulb memories of the assassination attempt on President Reagan. *Cognition, 16*(1), 63–80. https://doi.org/10.1016/0010-0277(84)90036-2

Pillemer, D. B., & Kuwabara, K. J. (2012). Directive functions of autobiographical memory: Theory and method. In D. C. Rubin & D. Berntsen (Eds.), *Understanding autobiographical memory: Theories and approaches* (pp. 181–202). Cambridge University Press. https://doi.org/10.1017/CBO9781139021937.014

Rathore, F., & Farooq, F. (2020). Information overload and infodemic in the COVID-19 pandemic. *Journal of the Pakistan Medical Association, 70*(5), S162–S165. https://doi.org/10.5455/JPMA.38

Rossi-Arnaud, C., Mastroberardino, S., Spataro, P., Santirocchi, A., Alessi, F., Naser, A., et al. (2023). Autobiographical memory in contact tracing: Evidence from the COVID-19 pandemic. *Frontiers in Psychology, 14*, 1244568. https://doi.org/10.3389/fpsyg.2023.1244568

Rouhani, N., Stanley, D., COVID-Dynamic Team, & Adolphs, R. (2023). Collective events and individual affect shape autobiographical memory. *Proceedings of the National Academy of Sciences, 120*(29), e2221919120. https://doi.org/10.1073/pnas.222191912

Rubin, D. C., & Berntsen, D. (2003). Life scripts help to maintain autobiographical memories of highly positive, but not highly negative, events. *Memory & Cognition, 31*(1), 1–14. https://doi.org/10.3758/BF03196077

Rubin, D. C., & Kozin, M. (1984). Vivid memories. *Cognition, 16*(1), 81–95. https://doi.org/10.1016/0010-0277(84)90037-4

Sartori, G., Agosta, S., Zogmaister, C., Ferrara, S. D., & Castiello, U. (2008). How to accurately detect autobiographical events. *Psychological Science, 19*(8), 772–780. https://doi.org/10.1111/j.1467-9280.2008.02156.x

Sprengholz, P., Henkel, L., Böhm, R., & Betsch, C. (2023). Historical narratives about the COVID-19 pandemic are motivationally biased. *Nature, 623*(7987), 588–593. https://doi.org/10.1038/s41586-023-06674-5

Talarico, J. M., & Rubin, D. C. (2003). Confidence, not consistency, characterizes flashbulb memories. *Psychological Science, 14*(5), 455–461. https://doi.org/10.1111/1467-9280.02453

Talarico, J. M., & Rubin, D. C. (2007). Flashbulb memories are special after all; in phenomenology, not accuracy. *Applied Cognitive Psychology, 21*(5), 557–578. https://doi.org/10.1002/acp.1293

Talarico, J. M., Bohn, A., & Wessel, I. (2019). The role of event relevance and congruence to social groups in flashbulb memory formation. *Memory, 27*(7), 985–997. https://doi.org/10.1080/09658211.2019.1616097

Talarico, J. M., LaBar, K. S., & Rubin, D. C. (2004). Emotional intensity predicts autobiographical memory experience. *Memory & cognition, 32*(7), 1118–1132. https://doi.org/10.3758/BF03196886

Taurisano, P., Lanciano, T., Alfeo, F., Bisceglie, F., Monaco, A., Sbordone, F. L., Abbatantuono, C., Costadura, S., Losole, J., Ruggiero, G., Iachini, S., Vimercati, L., Vacca, A., De Caro, M. F., & Curci, A. (2022). The COVID-19 stress perceived on social distance and gender-based implications. *Frontiers in Psychology, 13*, e846097. https://doi.org/10.3389/fpsyg.2022.846097

Tekcan, A. İ., Ece, B., Gülgöz, S., & Er, N. (2003). Autobiographical and event memory for 9/11: Changes across one year. *Applied Cognitive Psychology, 17*(9), *1057*–1066. https://doi.org/10.1002/acp.985

Tinti, C., Schmidt, S., Sotgiu, I., Testa, S., & Curci, A. (2009). The role of importance/consequentiality appraisal in flashbulb memory formation: The case of the death of Pope John Paul II. *Applied Cognitive Psychology, 23*(2), 236–253. https://doi.org/10.1002/acp.1452

Topçu, M. N., & Hirst, W. (2024). When the personal and the collective intersects: Memory, future thinking, and perceived agency during the COVID-19 pandemic. *Journal of Experimental Psychology: General, 153*(9), 2258–2278. https://doi.org/10.1037/xge0001624

Tull, M. T., Edmonds, K. A., Scamaldo, K. M., Richmond, J. R., Rose, J. P., & Gratz, K. L. (2020). Psychological outcomes associated with stay-at-home orders and the perceived impact of COVID-19 on daily life. *Psychiatry Research, 289*, 113098. https://doi.org/10.1016/j.psychres.2020.113098

Weaver, C. A. (1993). Do you need a "flash" to form a flashbulb memory? *Journal of Experimental Psychology: General, 122*(1), 39–46. https://doi.org/10.1037/0096-3445.122.1.39

Wolf, T., & Nusser, L. (2022). Maintaining intimacy during the COVID-19 pandemic. *Applied Cognitive Psychology, 36*(4), 954–961. https://doi.org/10.1002/acp.3960

Wright, D. B., & Gaskell, G. D. (1995). Flashbulb memories: Conceptual and methodological issues. *Memory, 3*(1), 67–80. https://doi.org/10.1080/09658219508251497

Xuan, Y., Greenberg, M., Umanath, S., & Coane, J. H. (2024). Flashbulb and first-hand memories for campus closings due to COVID-19: Consistency and change. *Memory, 33*(1), 84–99. https://doi.org/10.1080/09658211.2024.2404499

Yang, P., Qi, J., Zhang, S., Wang, X., Bi, G., Yang, Y., Sheng, B., & Yang, G. (2020). Feasibility study of mitigation and suppression strategies for controlling COVID-19 outbreaks in London and Wuhan. *PLoS One, 15*(8), e0236857. https://doi.org/10.1371/journal.pone.0236857

Beyond the Virus: The Pandemic of False Memories

Alaitz Aizpurua Sanz and Malen Migueles Seco

Abstract During the COVID-19 pandemic, there was not only a global health crisis, but also an overload of information, much of it erroneous or malicious, which favored the emergence of a parallel pandemic of false memories. Memory, understood as a reconstructive process, was particularly vulnerable to the effects of misinformation, cognitive biases, and emotional and social context. The massive dissemination of false news, both by mistake (misinformation) and intentionally (disinformation), was intensified by the use of social networks, the repetition of content, and the trust in sources, which increased the acceptance of inaccurate information as true. In addition, individual factors such as age, thinking style, actual knowledge about COVID-19, or emotional factors such as anxiety or depression influenced the propensity to recall and share false information. Although many of these false memories reveal the fragility of memory, others played an adaptive role by helping individuals reinterpret the past more positively, recall pleasant experiences, or envision a near future with better prospects, thus supporting psychological balance and well-being. Finally, the need to promote critical thinking and media literacy as prevention mechanisms is highlighted.

Keywords COVID-19 · Pandemic · False memories · Misinformation · Positivity

1 Introduction

During the years of the COVID-19 pandemic, not only did a virus spread uncontrollably, but there was also an avalanche of news, images, and contradictory narratives that deeply affected our memory of the events and influenced our behavior and decision-making. This chapter starts from the widely accepted premise that memory

A. Aizpurua Sanz (✉) · M. Migueles Seco
Faculty of Psychology, University of the Basque Country (UPV/EHU), Donostia-San Sebastian, Spain
e-mail: alaitz.aizpurua@ehu.eus; malen.migueles@ehu.eus

M. Migueles Seco, A. Aizpurua Sanz (eds.), *Autobiographical Memory and the Covid-19 Pandemic*, https://doi.org/10.1007/978-3-032-03895-1_6

is a reconstructive process subject to errors and distortions. First, it clarifies concepts related to false memories and explores how, why, and with what consequences they were generated in a pandemic context marked by fear, uncertainty, and widespread disinformation. It also analyzes the role of the media and social networks in a context where the desire to stay informed, the overabundance of content, the manipulation of information to bias public opinion, and conspiracy theories were constant. Individual differences in the formation of false memories during the COVID-19 pandemic will also be examined, given that cognitive, emotional, and age-related factors make us more or less susceptible. Next, various strategies will be proposed to reduce the impact of misinformation and improve our ability to assess the accuracy of information. The chapter also reflects on the adaptive function of the distortions and cognitive biases we developed during the pandemic, which helped us cope with the situation and improve our sense of well-being. A good example is positivity—the tendency to remember positive content and/or to modify negative information to make it more neutral or positive. Finally, some conclusions will be drawn, and several future perspectives will be proposed.

2 Memory as a Reconstructive Process

In the context of the COVID-19 pandemic, human memory revealed its most reconstructive nature. It is well-known that memory is not an exact record of our experiences, and that each act or attempt to retrieve a memory transforms the original episode. Bartlett (1932) already suggested that remembering is not about retrieving an exact copy of information, but an active process of reconstruction. The mind (re) organizes information and fills in gaps with inferences based on the context, prior knowledge, cultural content, or lived experiences. Thus, the reconstructive nature of memory involves distortions, the inclusion of errors, or the acceptance of content that—although logical, coherent, and therefore plausible—never actually happened.

Different memory errors were identified and categorized by Schacter in his influential book *The seven sins of memory: How the mind forgets and remembers* (2001), where he outlines seven "sins" of memory which, like the seven deadly sins, have both adaptive aspects and vulnerabilities that reveal memory's fragility (see also Schacter, 1999, 2022). On the one hand, there are errors of omission, such as *transience*, where information is lost over time; *blocking*, where known information is temporarily inaccessible; and *absent-mindedness*, where a lack of attention or multitasking leads to poor encoding of information. On the other hand, errors of commission occur when memory adds, alters, or distorts information. Schacter identifies three main types: *suggestibility*, which involves incorporating external information into our memories; *misattribution*, where we mistakenly attribute the origin or source of information; and *bias*, which arises when our beliefs, emotions, and experiences influence and shape our memories. Finally, *persistence* refers to the inability to forget memories or events we would rather not remember.

Recalling with great confidence events that never happened, or remembering them inaccurately with distorted details, is a common experience that highlights the malleability of memory. These errors can stem from external influences such as suggestion or social pressure, or from internal factors like biases rooted in our beliefs, expectations, or prior knowledge. Research on suggestion and post-event information has shown how easily external content can be incorporated into our memory of an event (Cheng et al., 2021; Loftus et al., 1978; Loftus, 2005; Frenda et al., 2011)—for example, remembering a non-existent stop sign at a car accident (Loftus et al., 1978). Likewise, autobiographical memories can be implanted simply by presenting true experiences along with a plausible but false one, such as getting lost in a shopping mall (Loftus & Pickrell, 1995), or meeting Bugs Bunny at Disneyland (Braun et al., 2002). Similarly, the use of doctored photos and videos can lead us to recall events we never personally experienced (Murphy et al., 2023; Nash et al., 2009) or vividly relive emotionally charged childhood episodes, such as a hypothetical hot-air balloon ride (Wade et al., 2002).

Our memory can also be distorted not only by misleading information or suggestion, but also by misjudging the source of our memories. Prior knowledge, personal experiences, and ideology can bias our interpretation of events and lead to the formation of false memories (Armaly & Enders, 2022; García-Bajos & Migueles, 2003; García-Bajos et al., 2012). For instance, García-Bajos et al. (2012; see also García-Bajos & Migueles, 2003) found that activating prior knowledge biased participants' recollection of a bank robbery shown in a video, leading them to remember stereotypical but false actions. Along similar lines, political identity can shape not only our perceptions but also our memories (Calvillo & Smelter, 2021; Calvillo et al., 2023; Frenda et al., 2013; Greene et al., 2021). For example, Greene et al. (2021) found that many voters in the UK who supported the Brexit referendum falsely remembered fake news that aligned with their political stance. These examples demonstrate that memory is a process that surprises us with its capacity and flexibility, but also with its fragility and ease of distortion, leading to false memories.

But why do these false memories arise? Research has proposed various theoretical models to help explain how people can remember events that never happened or distort meaningful aspects of real experiences. One such model is the *Constructive Memory Framework*, which suggests that memory does not function as a literal record of the past but as a reconstructive system that combines real information with inferences, expectations, and prior knowledge (Schacter, 1999). This constructive nature helps explain why, during the COVID-19 pandemic, many people recalled events that did not actually occur as they remembered them—for instance, believing they followed certain restrictions before they were officially imposed. Another model, the *Fuzzy Trace Theory* (Brainerd & Reyna, 2002), proposes that people store two types of memory traces simultaneously: one that captures specific details (*verbatim*) and another that captures the general meaning (*gist*) of the event. In contexts like the pandemic—possibly due to repeated experiences—people tend to rely more heavily on gist memories, increasing the risk of false memories. Finally, *Reality Monitoring Theory* (Johnson et al., 1993) offers a crucial component for understanding why false memories can feel so real. It suggests that when signals are ambiguous or faint—as happens under stress, fatigue, or repeated exposure to external images—people may fail to accurately judge the source of a memory and misattribute it to a personal experience, something imagined, or something merely suggested. For example, someone might recall having watched the Prime Minister's announcement of a state of emergency live, when in fact they only saw it later on the news or social media. These models help us understand how the pressure, overwhelming flow of information, and intense emotional load of the pandemic created a fertile ground for distorted memories.

3 False Memories Formation During the COVID-19 Pandemic

During the pandemic, the media and social networks became a breeding ground for the creation and spread of false information, revealing the fragility of memory. At this point, it is relevant to clarify concepts and distinguish between different types of misleading information. One criterion for distinction is the intent to deceive when people distribute and share false information (Pennycook & Rand, 2022). Based on this, Wardle and Derakhshan (2017) distinguish between *misinformation* and *disinformation*: misinformation refers to the sharing of false information without the conscious or deliberate intention to deceive or cause harm. For example, someone may innocently share with friends, family, or on social media that COVID-19 vaccines alter DNA or that drinking hot water on an empty stomach prevents infection—based on something read in a blog—without knowing the information is incorrect. Disinformation, by contrast, occurs when false information is intentionally created to mislead or manipulate. For instance, during the pandemic, organized campaigns with political or ideological agendas aimed to influence public opinion, manipulate the population's views, and discredit certain countries or institutions. This included conspiracy theories about the virus's origin, the inefficacy of medical services, vaccine side effects, and alternative treatment rumors.

Some researchers argue that misleading information is not always driven by an intent to deceive. In many cases, it is shared as a form of satire, humor, or entertainment (Tandoc et al., 2018). *Fake news* (Gelfert, 2018, 2021), meanwhile, is a subtype of disinformation, typically presented in the style of a news article or headline to increase its credibility. An example would be: "The WHO admits COVID-19 was intentionally developed and spread. It has also been suggested that *misinformation* has become an umbrella term referring to the spread of false or inaccurate information across various forms (e.g., social media, newspapers, personal comments, etc.) (Loftus & Klemfuss, 2024). In any case, both misinformation and disinformation can lead people to accept false content as factual, with significant personal and societal consequences.

Loftus (Greenspan & Loftus, 2021; Loftus & Klemfuss, 2024) raises an important question: Can everything we know about misleading information be applied to the spread of fake news on social media? In a classic post-event misinformation experiment, participants witness an event (e.g., a video or slides), and later, misleading information is introduced through open-ended questions, narratives, or confederates. In a final phase, memory for the event is assessed to determine whether false details have contaminated it (see Loftus, 2005, for a review). A consistent finding is that people tend to incorporate false or altered details into their memories, demonstrating that memory is susceptible to external influences. Research shows that the impact of misleading information is greater when the false information is repeated (Foster et al., 2012; Loftus, 2005), it is plausible and consistent with prior beliefs (Frenda et al., 2013; Schincariol et al., 2024), or it comes from a credible and prestigious source (Braun et al., 2002; Loftus, 2005; Pennycook & Rand, 2021). These findings were also observed during the pandemic. Greene and Murphy (2020) presented 3746 participants with six news items about COVID-19, two of which were false. A total of 22.5% recalled at least one false news item as true.

In 2020, the term *infodemic* was coined, defined by the WHO as an overabundance of information, some of which is accurate and some of which is false, that makes it difficult to distinguish facts from rumors (Zaracostas, 2020). The sheer volume of false content and rumours circulating online during the pandemic was astonishing, as it misled, confused, and manipulated people. For example, Islam et al. (2020) analyzed 2276 pieces of online content about COVID-19 and found that 82% contained misleading information. Cheng et al. (2021) identified and classified more than 6800 false rumors, many originating from social media. This material can serve as a valuable tool for researching disinformation, understanding how rumors spread, and designing strategies to combat and mitigate harmful misinformation. Like a virus, information spread rapidly through digital networks and platforms, generating a "pandemic of false memories." Factors such as uncertainty, constant repetition, social isolation, and ideological polarization created a fertile environment for distortions to flourish. The pandemic thus became an ideal context to study how misleading information interacts with our reconstructive memory processes.

4 False Memories: An Adaptive Process in the Face of Adversity

Referring to the adaptive nature and survival function of memory, some false memories generated during the pandemic may have served as a strategy to help maintain personal balance, cope with stress and anxiety, and improve mood and well-being. It is reasonable to assume that during the pandemic—considered a *chronic stressor* (Vinkers et al., 2020) due to the constant flow of bad news, deaths, and socioeconomic problems—our minds were overwhelmed with negative memories and future scenarios, which in turn reduced our sense of well-being and personal stability. This idea is consistent with the cognitive theory of depression (Beck, 2002) and the hopelessness theory (Abramson et al., 1989; Liu et al., 2015), which suggest that repeated exposure to aversive stimuli can lead to a perceived lack of control, the belief that we cannot escape the situation, and the expectation that our negative thoughts will inevitably come true. These theories are also aligned with the *mood-congruency effect*, which proposes that we tend to process, retain, and recall information that matches our current emotional state (Bower, 1981; Drace, 2013; Faul & Labar, 2023).

However, memory often comes to the rescue, encouraging us to avoid dwelling on misfortune, economic hardship, illness, or death, and instead to visualize the future more positively to face adversity (Ballance et al., 2022; Lalla & Sheldon, 2021; Szpunar et al., 2012). Nostalgia for positive past experiences may distort reality, but it helps us cope with present adversity and envision the future with greater resilience (Faul & De Brigard, 2022; Gammon & Ramshaw, 2021). Additionally, a phenomenon known as *unrealistic optimism* may arise—a cognitive bias that leads us to believe our future expectations are more favorable for ourselves and our loved ones than for others (Salgado & Berntsen, 2021; Weinstein, 1980). Similarly, remembering past experiences of self-efficacy has been shown to reduce fear related to COVID-19. Participants who recalled moments when they felt effective experienced a reduction in fear levels, suggesting that positive memories can serve as a tool for emotional well-being during stressful situations (Bakke et al., 2023; Zepeda et al., 2023). It is important to remember that one of the functions of autobiographical memory is emotional self-regulation (Conway, 2005; Conway & Pleydell-Pearce, 2000). In this way, memory modifies and distorts reality to be remembered more positively than it truly was, thereby promoting psychological well-being by reducing unpleasant emotions such as fear, anxiety, or sadness. For example, during lockdown, many people experienced time perception distortions and had trouble distinguishing one day from another due to routine and the absence of new experiences (Cellini et al., 2020; Ogden, 2020). In retrospect, they evaluated that period as being shorter in duration (Leitner et al., 2021). Some even perceived it as a relatively benign time (De la Fuente et al., 2021) or as a period of personal growth and resilience (Shigemura et al., 2022). In other words, reinterpreting the past may have helped people cope more adaptively.

A systematically observed finding during the pandemic is the *positivity bias*—that is, the tendency or preference for individuals to pay more attention to, process, and remember positive content more than negative. *The positivity bias has been observed in autobiographical memory and episodic future thinking during the COVID-19 health crisis* (Aizpurua et al., 2021; Lalla & Sheldon, 2021; Migueles & Aizpurua, 2024; Niziurski & Schaper, 2021; Salgado & Berntsen, 2021). For example, Lalla and Sheldon (2021) asked participants to generate specific past events and future projections related to COVID-19 in response to both positive and negative cues, finding that future-oriented events tended to be more positive than past ones. Similarly, Niziurski and Schaper (2021) observed that when participants recalled and anticipated pandemic-related events, positive events were rated as more emotional and vivid than negative ones. Other studies on *future thinking* also found a preference for positivity and a tendency to imagine a more hopeful future. For example, Aizpurua et al. (2021) presented 28 statements related to the pandemic and future outlook, half of them positive (e.g., "*I think this pandemic will strengthen us*") and the other half negative (e.g., "*I think this virus will mutate and we won't be able to beat it*"). *They found that participants not only agreed more with the positive than with the negative content, but also, in the recall task—and despite the acute global crisis at the time—they retrieved positive and negative thoughts in equal proportion. Moreover, reinforcing the idea of positivizing the situation, some participants even biased their recall by transforming originally negative content into more neutral or even positive ideas.* Similar findings, pointing to a tendency toward *rosy future simulation* (Szpunar et al., 2012), were reported in the study by Migueles and Aizpurua (2024). They explored how individuals conceptualized and recalled thoughts about the future during the COVID-19 lockdown, considering both the emotional valence (positive or negative) and the nature (personal or collective) of those thoughts. The results showed a clear positivity bias in personal future thoughts and, interestingly, a negativity bias in collective thoughts, which is consistent with the widely documented *domain-by-valence interaction* (Shrikanth et al., 2018; Shrikanth & Szpunar, 2021; Szpunar & Liu, 2023).

5 Individual Differences in the Formation of False Memories

Are we all equally vulnerable to false memories? The evidence suggests we are not. Individual differences explain why some people are more susceptible than others to accepting and retaining inaccurate information. These differences are related, among other factors, to ideological, emotional, and cultural variables. Various studies have shown how beliefs, thinking style, and the level of objective knowledge influenced susceptibility to false memories during the COVID-19 pandemic (Greene et al., 2022; Mangiulli et al., 2022). For example, in the early stages of vaccine development in 2020, rumors about their alleged dangers circulated, fueled by political tensions and advocacy groups. In this context, individuals who believed in conspiracy theories about vaccines were more likely to accept and recall false information on the subject (Mangiulli et al., 2022). Even those with a favorable attitude toward vaccination could have their trust undermined when exposed to misleading messages. Greene et al. (2022) showed that fake news consistent with participants' prior beliefs was more easily remembered, highlighting the role of ideological congruence as a memory bias. This example illustrates how individual differences in ideology or institutional trust not only influence the formation of false memories but also directly impact public decisions and collective health.

It has also been observed that people with an analytical thinking style—characterized by deliberate and critical reasoning—are less susceptible to false memories than those who rely on more intuitive, fast, and automatic thinking (Mangiulli et al., 2022; Schincariol et al., 2024). For instance, Mangiulli et al. (2022) conducted two studies, showing that between 41% and 55% of participants developed false memories after reading fake news about COVID-19. The findings revealed that analytical thinking was negatively correlated with vulnerability to these false narratives. In addition to analytical thinking, having a high level of *objective knowledge* about COVID-19 also reduced susceptibility to fake news. In contrast, those with high *subjective confidence* but limited real knowledge were more prone to accept, remember, and spread false information (Greene & Murphy, 2020; Scuotto et al., 2021).

Emotional factors such as anxiety and depression also played a central role in the formation of false memories. In stressful situations, individuals with high anxiety levels tended to intensify the emotional weight of their memories and showed greater susceptibility to misinformation because anxiety promotes *rapid intuitive thinking*, which facilitates the incorporation of inaccurate details into episodic memory (Greene & Murphy, 2020; Luminet et al., 2021). Furthermore, high anxiety levels were positively correlated with increased vividness and confidence in false memories, especially when the news was alarming or perceived as threatening (Scuotto et al., 2021). Depression, on the other hand, has been linked to a negativity bias, less specific recall, and *overgeneralized memory* (Hallford et al., 2022) features that increase the likelihood of integrating false information (Dalgleish & Werner-Seidler, 2014). Thus, people who experienced high anxiety or developed depression during the pandemic were more susceptible to false memories. These findings suggest that the emotional impact of the pandemic and negative emotions altered the mechanisms of encoding and retrieval of information.

Research also suggests that older adults are more susceptible to false memories due to a combination of cognitive and affective factors. Aging is associated with slower processing speed (Salthouse, 1996, 2019), reduced attentional capacity and limitations in encoding and retrieval processes (Nyberg et al., 2023), as well as increased reliance on heuristics and preexisting schemas (García-Bajos et al., 2012; Schacter et al., 2011), which facilitates the integration of distorted or erroneous elements. But does this mean that older adults were more susceptible than younger people to misinformation or misleading content shared on social media during the pandemic? Young people aged 16–30 primarily obtained their information through the internet and social networks and often lacked the skills to detect false information or cross-check it with other sources (Ofcom, 2024). As younger people relied more on the internet for health-related information and possessed less knowledge overall, it is reasonable to think that they may have been more vulnerable to misinformation and disinformation about COVID-19 and the vaccines (Nugent et al., 2025). In a meta-analysis by Sultan et al. (2024), 31 studies were reviewed to examine demographic factors related to susceptibility to misleading information. Researchers analyzed participants' ability to distinguish between true and false news and their *response bias*—that is, their tendency to label news as true or false. The results showed that older adults were more cautious when evaluating the truthfulness of information and better discriminated between true and false news (see also Brashier & Schacter, 2020). However, other studies have shown that older adults' ability to discriminate genuine from manipulated images of real-world scenes is slightly poorer than younger and middle-aged adults (Nightingale et al.,

2022), and that adults over 65 are more likely to share false or unverified news on social media than younger individuals (Guess et al., 2019; Matchanova et al., 2023).

Another area where age-related differences are clearly evident is in the *positivity bias*. Older adults tend to show a stronger preference for positive events compared to younger individuals (Aizpurua et al., 2021; García-Bajos et al., 2017; Mather & Carstensen, 2005), resulting not only in a greater proportion of positive memories but also a greater tendency to reinterpret negative experiences in a more neutral or even favorable light. This phenomenon, known as the *positivity effect*, has been consistently observed across a wide range of stimuli, including words (Kensinger, 2008; Hamilton & Allard, 2020), scenes, drawings, and faces (e.g., Charles et al., 2003; Mather & Carstensen, 2005; Reed et al., 2014; Mammarella et al., 2016), as well as autobiographical memories and simulations of future events (Gallo et al., 2011; Cole et al., 2016; García-Bajos et al., 2017; Salgado & Berntsen, 2020; Szpunar et al., 2012). This effect has also been documented in highly emotional or uncertain situations, such as the COVID-19 pandemic—particularly during moments like the declaration of the state of emergency or the nationwide lockdown (Aizpurua et al., 2021; Carstensen et al., 2020). For example, Carstensen et al. (2020) asked participants aged 18–76 to report the frequency and emotional valence of their experiences, as well as their fear of infection and potential health complications. Older adults reported greater positivity, demonstrated better emotional regulation, and indicated higher levels of well-being compared to younger adults. Similarly, in the aforementioned study by Aizpurua et al. (2021), when young, middle-aged and older participants were presented with both positive and negative COVID-related statements, younger adults recalled a greater number of items overall, but older adults recalled a higher proportion of positive content. Moreover, both older and middle-aged participants were more likely than younger adults to reinterpret negative statements more neutrally or positively. One of the main theoretical explanations for this effect stems from *Socioemotional Selectivity Theory* (Carstensen, 2006; Carstensen et al., 1999; Reed & Carstensen, 2012), which proposes that the awareness of life's temporal limitations—a hallmark of aging—leads to a motivational shift. As individuals age, they increasingly prioritize emotionally meaningful goals and positive experiences, which in turn foster psychological balance and a greater sense of well-being.

6 Is It Possible to Reduce the Impact of False Information?

People are generally unable to identify false information or resist suggestions spontaneously. The data are clear: exposure to misinformation compromises memory accuracy. Even when we are aware that the information is false or misleading, its effects are difficult to mitigate, and we continue to rely on it in our reasoning, thinking, and behavior (Lewandowsky et al., 2017; Van Huijstee et al., 2022). This may be because false information is perceived as familiar, aligns with our existing beliefs, or has already been integrated into our knowledge. For example, van

Huijstee et al. (2022) investigated how false information affects attitudes, even after it has been corrected. Participants in their study read fake news (positive or negative) about COVID-19 recovery or death rates at a fictional hospital and, even after the misinformation was corrected, the negative fake news continued to influence their perception of the hospital.

Due to its real-world relevance, strategies have been studied to minimize the impact of misinformation on memory. While not foolproof, some techniques have shown promise—such as providing explicit warnings that information may be false (see Blank & Launay, 2014, for a meta-analysis), or asking participants to actively detect discrepancies or changes in information (Bayley et al., 2021; Butler & Loftus, 2018). While many of these approaches may not directly apply to everyday situations or high-impact events like the COVID-19 pandemic, they help us understand the underlying cognitive processes. During the pandemic, false information was often presented and spread by seemingly credible sources, such as medical associations, politicians, or influencers with a large number of followers. These messages were often consistent with the current situation and were repeated frequently, increasing the perceived credibility of the content (Greene & Murphy, 2020; Unkelbach & Speckmann, 2021). The public's need to stay informed and to share news as it emerged contributed to a lack of critical thinking and difficulty identifying inconsistencies or errors. Moreover, the effects of misinformation take on a new dimension when we consider that current technology enables the generation and manipulation of photos and videos (e.g., deepfakes). It has become easy to create AI-generated audiovisual content in which people appear to say or do things they never actually did (Pataranutaporn et al., 2023). However, the same technology that biases and deceives us could also help reduce the impact of misinformation on social media. For example, modifying social media algorithms to include prompts or warnings when users are about to share information—indicating that the content may be false, outdated, or from an unreliable source—could encourage greater caution (Frenda et al., 2011; Zhu et al., 2012). If people receive such alerts while browsing or sharing on social media, they might be more hesitant to spread questionable content.

Several authors argue that people often do not prioritize accuracy when processing information on social media. This "truth insensitivity," or difficulty distinguishing between true and false information, can be reduced through approaches that encourage reflective thinking before sharing content (Fazio, 2020; Pennycook et al., 2020). In fact, it may be simple and quick to promote and teach strategies that help people recognize misleading information (Pennycook et al., 2021; van der Linden et al., 2021). Effective strategies should aim to enhance people's ability to distinguish between true and false information, mitigate the impact of belief biases, consider the accuracy of content when making decisions, and ultimately promote media literacy. In this vein, Pennycook and colleagues (Pennycook et al., 2021; Pennycook & Rand, 2022) developed a model called the inattention-based account of misinformation sharing, which focuses on users' lack of attention to accuracy. They argue that misinformation is not primarily due to users' ideology, lack of critical thinking, or inability to distinguish truth from falsehood but rather due to the design of social media platforms, which prioritize other factors, such as speed, emotional appeal, perceived relevance, or the popularity of the source (likes, followers). Shifting users' attention toward the accuracy of the information they consume and share could be key to reducing the spread of misinformation among users who do not intend to deceive others. These authors introduce the concept of "nudges"—small prompts that encourage users to reflect on the accuracy of what they share. For example, simply asking users to evaluate a news item's truthfulness and check the facts before sharing increases the quality of the information that circulates. In short, devoting attention to thinking about accuracy, or detecting discrepancies and inaccuracies, may be an effective strategy to curb the spread of false information.

7 Conclusions

The COVID-19 pandemic brought not only an unprecedented global health crisis but also put our cognitive processes—particularly memory—to the test. In a context marked by fear, information overload, and uncertainty, people were especially vulnerable to generating distorted or outright false memories. These false memories were not mere cognitive errors; they were deeply rooted in emotional, social, and cultural processes. However, false memories do not always imply dysfunction. In many cases during the pandemic, they served an adaptive function: helping people reinterpret the past more positively, minimize suffering, regain a sense of control, or maintain a more hopeful view of the future. This phenomenon underscores the reconstructive nature of memory and its key role in emotional self-regulation and survival.

The role of the media—and particularly social networks—was crucial in both spreading useful information and amplifying misinformation. In many cases, people shared false content without verifying its truthfulness, driven by the desire to stay informed and influenced by ideological alignment or trust in certain sources. This dynamic was intensified by the constant repetition of certain messages, which contributed to reinforcing their perceived credibility. In this scenario, it is crucial to promote greater media literacy and foster critical thinking as essential tools to mitigate the harmful effects of misinformation. The ability to question the truthfulness of content and recognize cognitive biases can significantly reduce the acceptance and dissemination of false information. Likewise, incorporating warnings and automatic filters on social networks—designed to alert users to the possible inaccuracy of content before they share it—could be an effective strategy. These small prompts or nudges may help reduce the uncritical spread of misleading content.

We can conclude that everyone is susceptible to false memories, but the factors explaining this vulnerability vary depending on the individual's profile. Among the cognitive variables, conspiratorial beliefs, intuitive thinking styles, and overconfidence without actual knowledge contributed to the acceptance and dissemination of misinformation. On an emotional level, anxiety and depression increased susceptibility to memory distortions, affecting the emotional intensity, specificity, and reconstruction of memories. Regarding age, while older adults present certain cognitive limitations and a pronounced positivity bias that may facilitate memory errors, they were also found to be more cautious in distinguishing false news. Younger adults, who are more exposed to the internet and sometimes lack critical thinking capacity, were also particularly vulnerable to COVID-19-related misinformation. Thus, age alone does not determine susceptibility; rather, it interacts with other cognitive, emotional, and contextual variables.

Looking ahead, it is crucial to continue researching the role of false memories not only as distortions but also as potential adaptive psychological mechanisms. Understanding when such memories are harmful and to what extent they may be protective will be key to addressing their implications for mental health and social life. Another essential line of work involves implementing educational programs that promote media literacy and analytical thinking from an early age. Training in identifying reliable sources, questioning what one reads, and developing critical thinking skills could serve as powerful defences in future misinformation crises. It would also be valuable to conduct longitudinal studies to analyze how individual and collective memories of the pandemic evolve over time, what factors influence their modification, and how they are reconfigured in response to changing social or political contexts. This final research line could provide insight into how history is rewritten through memory.

Acknowledgment The authors acknowledge the financial support from the Basque Government / Eusko Jaurlaritza. Grant Number: GIU21/70. General Funding for Research Groups, Award Number: IT1604/22.

References

Abramson, L. Y., Metalsky, G. I., & Alloy, L. B. (1989). Hopelessness depression: A theory-based subtype of depression. *Psychological Review, 96*(2), 358–372. https://doi.org/10.1037/0033-295X.96.2.358

Aizpurua, A., Migueles, M., & Aranberri, M. (2021). Age differences in memory and positivity bias during the COVID-19 lockdown. *Journal of Applied Research in Memory and Cognition, 10*(4), 504–514. https://doi.org/10.1016/j.jarmac.2021.06.005

Armaly, M. T., & Enders, A. M. (2022). Filling in the gaps: False memories and partisan bias. *Political Psychology, 44*(2), 281–299. https://doi.org/10.1111/pops.12841

Bakke, B. T., Beaty, R. E., Silvia, P. J., & Silvia, P. (2023). Autobiographical memory for self-efficacy reduces COVID-19 related fear. *Memory, 31*(1), 21–29. https://doi.org/10.1080/09658211.2022.2142544

Ballance, S., Charlesworth, T. E. S., & Gureckis, T. M. (2022). Nostalgic memories promote optimism. *Emotion, 22*(6), 1157–1169. https://doi.org/10.1037/emo0001106

Bartlett, F. C. (1932). *Remembering: A study in experimental and social psychology*. Cambridge University Press.

Bayley, P. J., Brown, A. D., Castro, C. A., & Glenn, D. E. (2021). Enhancing resistance to misinformation with attentional strategies. *Journal of Applied Research in Memory and Cognition, 10*(3), 366–374. https://doi.org/10.1016/j.jarmac.2021.05.007

Beck, A. T. (2002). Cognitive models of depression. In R. L. Leahy & E. T. Dowd (Eds.), *Clinical advances in cognitive psychotherapy: Theory and application* (pp. 29–61). Springer.

Blank, H., & Launay, C. (2014). How to protect eyewitness memory against the misinformation effect: A meta-analysis of post-warning studies. *Journal of Applied Research in Memory and Cognition, 3*(2), 77–88. https://doi.org/10.1016/j.jarmac.2014.03.005

Bower, G. H. (1981). Mood and memory. *American Psychologist, 36*(2), 129–148. https://doi.org/10.1037/0003-066X.36.2.129

Brainerd, C. J., & Reyna, V. F. (2002). Fuzzy-trace theory and false memory. *Current Directions in Psychological Science, 11*(5), 164–169. https://doi.org/10.1111/1467-8721.00

Brashier, N. M., & Schacter, D. L. (2020). Aging in an era of fake news. *Current Directions in Psychological Science, 29*(3), 316–323. https://doi.org/10.1177/0963721420915872

Braun, K. A., Ellis, R., & Loftus, E. F. (2002). Make my memory: How advertising can change our memories of the past. *Psychology & Marketing, 19*(1), 1–23. https://doi.org/10.1002/mar.1000

Butler, A. C., & Loftus, E. F. (2018). The misinformation effect: Past, present, and future. *Current Directions in Psychological Science, 27*(1), 20–25. https://doi.org/10.1177/0963721417738820

Calvillo, D. P., & Smelter, T. J. (2021). Political ideology predicts perceptions of the threat of COVID-19 (and susceptibility to fake news about it). *Social Psychological and Personality Science, 12*(8), 1523–1530. https://doi.org/10.1177/1948550620940539

Calvillo, D. P., Harris, J. D., & Hawkins, W. C. (2023). Partisan bias in false memories for misinformation about the 2021 US Capitol riot. *Memory, 31*(1), 137–146. https://doi.org/10.1080/09658211.2022.2127771

Carstensen, L. L. (2006). The influence of a sense of time on human development. *Science, 312*(5782), 1913–1915. https://doi.org/10.1126/science.1127488

Carstensen, L. L., Isaacowitz, D. M., & Charles, S. T. (1999). Taking time seriously: A theory of socioemotional selectivity. *American Psychologist, 54*(3), 165–181. https://doi.org/10.1037/0003-066X.54.3.165

Carstensen, L. L., Shavit, Y. Z., & Barnes, J. T. (2020). Age advantages in emotional experience persist even under threat from the COVID-19 pandemic. *Psychological Science, 31*(11), 1374–1385. https://doi.org/10.1177/0956797620967261

Cellini, N., Canale, N., Mioni, G., & Costa, S. (2020). Changes in sleep pattern, sense of time and digital media use during COVID-19 lockdown in Italy. *Journal of Sleep Research, 29*(4), e13074. https://doi.org/10.1111/jsr.13074

Charles, S. T., Mather, M., & Carstensen, L. L. (2003). Aging and emotional memory: The forgettable nature of negative images for older adults. *Journal Experimental of Psychology: General, 132*, 310–324. https://doi.org/10.1037/0096-3445.132.2.310

Cheng, G., Ma, Y., Li, X., & Ma, S. (2021). A database of COVID-19 misinformation on the web. *Journal of Medical Internet Research, 23*(4), e26043. https://doi.org/10.2196/26043

Cole, S. N., Staugaard, S. R., & Berntsen, D. (2016). Inducing involuntary and voluntary mental time travel using a laboratory paradigm. *Memory & Cognition, 44*, 376–389. https://doi.org/10.3758/s13421-015-0564-9

Conway, M. A. (2005). Memory and the self. *Journal of Memory and Language, 53*(4), 594–628. https://doi.org/10.1016/j.jml.2005.08.005

Conway, M. A., & Pleydell-Pearce, C. W. (2000). The construction of autobiographical memories in the self-memory system. *Psychological Review, 107*(2), 261–288. https://doi.org/10.1037/0033-295X.107.2.261

Dalgleish, T., & Werner-Seidler, A. (2014). Disruptions in autobiographical memory processing in depression and the emergence of memory therapeutics. *Trends in Cognitive Sciences, 18*(11), 596–604. https://doi.org/10.1016/j.tics.2014.06.010

De la Fuente, J., Kauffman, D. F., & López-Ramos, V. M. (2021). COVID-19 as a stressor: Recovery of psychological well-being through self-regulation. *Journal of Clinical Medicine, 10*(4), 798. https://doi.org/10.3390/jcm10040798

Drace, K. R. (2013). Positive mood and memory: The role of encoding. *Cognition and Emotion, 27*(1), 95–106. https://doi.org/10.1080/02699931.2012.675885

Faul, L., & De Brigard, F. (2022). Counterfactual thinking and memory modification: Implications for well-being. *Current Opinion in Psychology, 43*, 215–220. https://doi.org/10.1016/j.copsyc.2021.07.023

Faul, L., & Labar, K. S. (2023). Mood-congruent memory revisited. *Psychological Review, 130*(6), 1421. https://doi.org/10.1037/rev0000394

Fazio, L. K. (2020). Pausing to consider why a headline is true or false can help reduce the sharing of false news. *The Harvard Kennedy School Misinformation Review, 1*(2). https://doi.org/10.37016/mr-2020-009

Foster, J. L., Huthwaite, T., Yesberg, J. A., Garry, M., & Loftus, E. F. (2012). Repetition, not number of sources, increases both susceptibility to misinformation and confidence in the accuracy of eyewitnesses. *Acta Psychologica, 139*(2), 320–326. https://doi.org/10.1016/j.actpsy.2011.12.004

Frenda, S. J., Knowles, E. D., Saletan, W., & Loftus, E. F. (2013). False memories of fabricated political events. *Journal of Experimental Social Psychology, 49*(2), 280–286. https://doi.org/10.1016/j.jesp.2012.10.013

Frenda, S. J., Nichols, R. M., & Loftus, E. F. (2011). Current issues and advances in misinformation research. *Current Directions in Psychological Science, 20*(1), 20–23. https://doi.org/10.1177/0963721410396620

Gallo, D. A., Korthauer, L. E., McDonough, I. M., Teshale, S., & Johnson, E. L. (2011). Age-related positivity effects and autobiographical memory detail: Evidence from a past/future source memory task. *Memory, 19*, 641–652. https://doi.org/10.1080/09658211.2011.595723

Gammon, S., & Ramshaw, G. (2021). Nostalgia, heritage, and the post-pandemic recovery. *Annals of Tourism Research Empirical Insights, 2*(2), 100024. https://doi.org/10.1016/j.annale.2021.100024

García-Bajos, E., & Migueles, M. (2003). False memories for script actions in a mugging account. *European Journal of Cognitive Psychology, 15*(2), 195–208. https://doi.org/10.1080/09541440244000102

García-Bajos, E., Migueles, M., & Aizpurua, A. (2017). Age-based positivity effects in imagining and recalling future positive and negative autobiographical events. *Frontiers in Psychology, 8*, 1700. https://doi.org/10.3389/fpsyg.2017.01700

García-Bajos, E., Migueles, M., & Aizpurua, A. (2012). Bias of script-driven processing on eyewitness memory in young and older adults. *Applied Cognitive Psychology, 26*(5), 737–745. https://doi.org/10.1002/acp.2854

Gelfert, A. (2018). Fake news: A definition. *Informal Logic, 38*(1), 84–117. https://doi.org/10.22329/il.v38i1.5068

Gelfert, A. (2021). Fake news, false beliefs, and the fallible art of knowledge maintenance. In S. Bernecker, A. K. Flowerree, & T. Grundmann (Eds.), *The epistemology of fake news* (pp. 310–333). Oxford University Press.

Greene, C. M., de Saint Laurent, C., Hegarty, K., & Murphy, G. (2022). False memories for true and false vaccination information form in line with pre-existing vaccine opinions. *Applied Cognitive Psychology, 36*(6), 1200–1208. https://doi.org/10.1002/acp.4002

Greene, C. M., & Murphy, G. (2020). Individual differences in susceptibility to false memories for COVID-19 fake news. *Cognitive Research: Principles and Implications, 5*, 63. https://doi.org/10.1186/s41235-020-00262-1

Greene, C. M., Nash, R. A., & Murphy, G. (2021). Misremembering Brexit: Partisan bias and individual predictors of false memories for fake news stories among Brexit voters. *Memory, 29*(5), 587–604. https://doi.org/10.1080/09658211.2021.1923754

Greenspan, R. L., & Loftus, E. F. (2021). Pandemics and infodemics: Research on the effects of misinformation on memory. *Human Behavior and Emerging Technologies, 3*(1), 8–12. https://doi.org/10.1002/hbe2.228

Guess, A., Nagler, J., & Tucker, J. (2019). Less than you think: Prevalence and predictors of fake news dissemination on Facebook. *Science Advances, 5*(1), eaau4586. https://doi.org/10.1126/sciadv.aau4586

Hallford, D. J., Rusanov, D., Yeow, J. J., & Barry, T. J. (2022). Reduced specificity and increased overgenerality of autobiographical memory persist as cognitive vulnerabilities in remitted major depression: A meta-analysis. *Clinical Psychology & Psychotherapy, 29*(5), 1515–1529. https://doi.org/10.1002/cpp.2786

Hamilton, L. J., & Allard, E. S. (2020). Words matter: Age-related positivity in episodic memory for abstract but not concrete words. *Aging, Neuropsychology, & Cognition, 27*, 595–616. https://doi.org/10.1080/13825585.2019.1657556

Islam, M. S., Sarkar, T., Khan, S. H., Mostofa Kamal, A.-H., Hasan, S. M. M., Kabir, A., Yeasmin, D., Islam, M. A., Chowdhury, K. I. A., Anwar, K. S., Chughtai, A. A., & Seale, H. (2020). COVID-19–related infodemic and its impact on public health: A global social media analysis. *The American Journal of Tropical Medicine and Hygiene, 103*(4), 1621–1629. https://doi.org/10.4269/ajtmh.20-0812

Johnson, M. K., Hashtroudi, S., & Lindsay, D. S. (1993). Source monitoring. *Psychological Bulletin, 114*(1), 3–28. https://doi.org/10.1037/0033-2909.114.1.3

Kensinger, E. A. (2008). Age differences in memory for arousing and nonarousing emotional words. *The Journals of Gerontology: Series B, 638*, 13–18. https://doi.org/10.1093/geronb/63.1.p13

Lalla, N. R., & Sheldon, K. M. (2021). Nostalgia prompts optimism and reduces stress by increasing meaning. *Memory, 29*(9), 1083–1090. https://doi.org/10.1080/09658211.2021.1952676

Leitner, M. C., Wessely, A. C., & Wührer, W. (2021). The subjective experience of time during the first COVID-19 lockdown in Austria: The role of affect, stress and social context. *PLoS One, 16*(9), e0256112. https://doi.org/10.1371/journal.pone.0256112

Lewandowsky, S., Ecker, U. K. H., & Cook, J. (2017). Beyond misinformation: Understanding and coping with the "post-truth" era. *Journal of Applied Research in Memory and Cognition, 6*(4), 353–369. https://doi.org/10.1016/j.jarmac.2017.07.008

Liu, R. T., Kleiman, E. M., Nestor, B. A., & Cheek, S. M. (2015). The hopelessness theory of depression: A quarter-century in review. *Clinical Psychology: Science and Practice, 22*(4), 345–365. https://doi.org/10.1111/cpsp.12125

Loftus, E. F. (2005). Planting misinformation in the human mind: A 30-year investigation of the malleability of memory. *Learning & Memory, 12*(4), 361–366. https://doi.org/10.1101/lm.94705

Loftus, E. F., & Klemfuss, J. Z. (2024). Misinformation–past, present, and future. *Psychology, Crime & Law, 30*(4), 312–318. https://doi.org/10.1080/1068316X.2023.2219813

Loftus, E. F., Miller, D. G., & Burns, H. J. (1978). Semantic integration of verbal information into a visual memory. *Journal of Experimental Psychology: Human Learning and Memory, 4*(1), 19–31. https://doi.org/10.1037/0278-7393.4.1.19

Loftus, E. F., & Pickrell, J. E. (1995). The formation of false memories. *Psychiatric Annals, 25*(12), 720–725.

Luminet, O., Nielson, K. A., & Ridout, N. (2021). Having no words for feelings: Alexithymia as a fundamental personality dimension at the interface of cognition and emotion. *Cognition and Emotion, 35*(3), 435–448. https://doi.org/10.1080/02699931.2021.1916442

Mammarella, N., Di Domenico, A., Palumbo, R., & Fairfield, B. (2016). When green is positive and red is negative: Aging and the influence of color on emotional memories. *Psychology and Aging, 31*, 914–926. https://doi.org/10.1037/pag000012

Mangiulli, I., Battista, F., Kafi, N. A., Coveliers, E., Webster, T. C., Curci, A., & Otgaar, H. (2022). False memory and COVID-19: How people fall for fake news about COVID-19 in digital contexts. *Frontiers in Psychology, 13*, 972004. https://doi.org/10.3389/fpsyg.2022.972004

Matchanova, A., Woods, S. P., Neighbors, C., Beltran-Najera, I., Alex, C., Johnson, B., et al. (2023). 7 Does neurocognition contribute to age-related differences in the accuracy and sharing of COVID-19 misinformation? *Journal of the International Neuropsychological Society, 29*(s1), 322–323. https://doi.org/10.1017/S1355617723004447

Mather, M., & Carstensen, L. L. (2005). Aging and motivated cognition: The positivity effect in attention and memory. *Trends in Cognitive Science, 9*, 496–502. https://doi.org/10.1016/j.tics.2005.08.005

Migueles, M., & Aizpurua, A. (2024). Mental journey to the future and memory for future thoughts during the COVID-19 pandemic lockdown. *Applied Cognitive Psychology, 38*(1), e4152. https://doi.org/10.1002/acp.4152

Murphy, G., Maher, J., Ballantyne, L., Barrett, E., Cowman, C. S., Dawson, C. A., et al. (2023). How do participants feel about the ethics of rich false memory studies? *Memory, 31*(4), 474–481. https://doi.org/10.1080/09658211.2023.2170417

Nash, R. A., Wade, K. A., & Lindsay, D. S. (2009). Digitally manipulating memory: Effects of doctored videos and imagination in distorting beliefs and memories. *Memory & Cognition, 37*(4), 494–503. https://doi.org/10.3758/MC.37.4.414

Nightingale, S. J., Wade, K. A., & Watson, D. G. (2022). Investigating age-related differences in ability to distinguish between original and manipulated images. *Psychology and Aging, 37*(3), 326–337. https://doi.org/10.1037/pag0000682

Niziurski, J. A., & Schaper, M. L. (2021). Psychological wellbeing, memories, and future thoughts during the COVID-19 pandemic. *Current Psychology, 1*, 2422–2435. https://doi.org/10.1007/s12144-021-01969-0

Nugent, G. R., Anselmi, D., & Chin, B. N. (2025). Social media, conspiracy beliefs, and covid-19 vaccines: A survey study of emerging and middle-aged adults in the United States. *Social Sciences, 14*(1), 34. https://doi.org/10.3390/socsci14010034

Nyberg, L., Andersson, M., Lundquist, A., Baaré, W. F., Bartrés-Faz, D., Bertram, L., et al. (2023). Individual differences in brain aging: Heterogeneity in cortico-hippocampal but not caudate atrophy rates. *Cerebral Cortex, 33*(9), 5075–5081. https://doi.org/10.1093/cercor/bhac400

Ofcom. (2024). *Adults' media use and attitudes*. Ofcom. https://www.ofcom.org.uk/research-and-data/media-literacy-research/adults/adults-media-use-and-attitudes

Ogden, R. S. (2020). The passage of time during the UK Covid-19 lockdown. *PLoS One, 15*(7), e0235871. https://doi.org/10.1371/journal.pone.0235871

Pataranutaporn, P., Liu, R., Finn, E., & Maes, P. (2023). Influencing human–AI interaction by priming beliefs about AI can increase perceived trustworthiness, empathy and effectiveness. *Nature Machine Intelligence, 5*(10), 1076–1086. https://doi.org/10.1038/s42256-023-00720-7

Pennycook, G., Epstein, Z., Mosleh, M., Arechar, A. A., Eckles, D., & Rand, D. G. (2021). Shifting attention to accuracy can reduce misinformation online. *Nature, 592*(7855), 590–595. https://doi.org/10.1038/s41586-021-03344-2

Pennycook, G., McPhetres, J., Zhang, Y., Lu, J. G., & Rand, D. G. (2020). Fighting COVID-19 misinformation on social media: Experimental evidence for a scalable accuracy-nudge intervention. *Psychological Science, 31*(7), 770–780. https://doi.org/10.1177/0956797620939054

Pennycook, G., & Rand, D. G. (2021). The psychology of fake news. *Trends in Cognitive Sciences, 25*(5), 388–402. https://doi.org/10.1016/j.tics.2021.02.007

Pennycook, G., & Rand, D. G. (2022). Accuracy prompts are a replicable and generalizable approach for reducing the spread of misinformation. *Nature Communications, 13*(1), 2333. https://doi.org/10.1038/s41467-022-30073-5

Reed, A., & Carstensen, L. (2012). The theory behind the age-related positivity effect. *Frontiers in Psychology, 3*, 339. https://doi.org/10.3389/fpsyg.2012.00339

Reed, A. E., Chan, L., & Mikels, J. A. (2014). Meta-analysis of the age-related positivity effect: Age differences in preferences for positive over negative information. *Psychology & Aging, 29*, 1–15. https://doi.org/10.1037/a0035194

Salgado, S., & Berntsen, D. (2020). My future is brighter than yours: The positivity bias in episodic future thinking and future self-images. *Psychological Research, 84*, 1829–1845. https://doi.org/10.1007/s00426-019-01189-z

Salgado, S., & Berntsen, D. (2021). Optimism bias during the COVID-19 pandemic: Empirical findings and theoretical implications. *Memory & Cognition, 49*(7), 1319–1332. https://doi.org/10.3758/s13421-021-01169-x

Salthouse, T. A. (1996). The processing-speed theory of adult age differences in cognition. *Psychological Review, 103*(3), 403–428. https://doi.org/10.1037/0033-295X.103.3.403

Salthouse, T. A. (2019). Trajectories of normal cognitive aging. *Psychology and Aging, 34*(1), 17–24. https://doi.org/10.1037/pag0000288

Schacter, D. L. (1999). The seven sins of memory: Insights from psychology and cognitive neuroscience. *American Psychologist, 54*(3), 182–203. https://doi.org/10.1037/0003-066X.54.3.182

Schacter, D. L. (2001). *The seven sins of memory: How the mind forgets and remembers*. Houghton Mifflin.

Schacter, D. L. (2022). Memory sins in applied settings: What kind of progress? *Journal of Applied Research in Memory and Cognition, 11*(4), 445–460. https://doi.org/10.1037/mac0000078

Schacter, D. L., Guerin, S. A., & Jacques, P. L. S. (2011). Memory distortion: An adaptive perspective. *Trends in Cognitive Sciences, 15*(10), 467–474. https://doi.org/10.1016/j.tics.2011.08.004

Schincariol, A., Otgaar, H., Greene, C. M., Murphy, G., Riesthuis, P., Mangiulli, I., & Battista, F. (2024). Fake memories: A meta-analysis on the effect of fake news on the creation of false memories and false beliefs. *Memory, Mind and Media, 3*, e17. https://doi.org/10.1017/mem.2024.14

Scuotto, V., Greene, C. M., & Murphy, G. (2021). False memory formation for COVID-19 fake news: The roles of subjective and objective knowledge, fear, and news literacy. *Cognitive Research: Principles and Implications, 6*(1), 5. https://doi.org/10.1186/s41235-021-00274-2

Shigemura, J., Kurokawa, S., Kurosawa, M., & Nomura, S. (2022). Mental health impact and perceived challenges of COVID-19 pandemic on university students in Japan: A cross-sectional survey. *BMC Psychiatry, 22*, 223. https://doi.org/10.1186/s12888-022-03879-4

Shrikanth, S., & Szpunar, K. K. (2021). The good old days and the bad old days: Evidence for a valence-based dissociation between personal and public memory. *Memory, 29*, 180–192. https://doi.org/10.1080/09658211.2020.1871024

Shrikanth, S., Szpunar, P. M., & Szpunar, K. K. (2018). Staying positive in a dystopian future: A novel dissociation between personal and collective cognition. *Journal of Experimental Psychology: General, 147*(8), 1200–1210. https://doi.org/10.1037/xge0000421

Sultan, M., Tump, A. N., Ehmann, N., Lorenz-Spreen, P., Hertwig, R., Gollwitzer, A., & Kurvers, R. H. (2024). Susceptibility to online misinformation: A systematic meta-analysis of demographic and psychological factors. *Proceedings of the National Academy of Sciences, 121*(47), e2409329121. https://doi.org/10.1073/pnas.2409329121

Szpunar, K. K., Addis, D. R., & Schacter, D. L. (2012). Memory for emotional simulations: Remembering a rosy future. *Psychological Science, 23*(1), 24–29. https://doi.org/10.1177/0956797611422237

Szpunar, K. K., & Liu, J. H. (2023). On the relations between personal and national event cognition: Theoretical and methodological considerations. *Journal of Applied Research in Memory and Cognition, 12*, 38–42. https://doi.org/10.1037/mac0000111

Tandoc, E. C., Lim, Z. W., & Ling, R. (2018). Defining "fake news": A typology of scholarly definitions. *Digital Journalism, 6*(2), 137–153. https://doi.org/10.1080/21670811.2017.1360143

Unkelbach, C., & Speckmann, F. (2021). Familiarity and the truth: When repetition increases perceived truth without affecting belief. *Memory, 29*(4), 478–488. https://doi.org/10.1080/09658211.2021.1911522

Van der Linden, S., Roozenbeek, J., Compton, J., & Maertens, R. (2021). Inoculating against fake news about COVID-19. *Frontiers in Psychology, 12*, 566790. https://doi.org/10.3389/fpsyg.2021.566790

Van Huijstee, D., Vermeulen, I., Kerkhof, P., & Droog, E. (2022). Continued influence of misinformation in times of COVID-19. *International Journal of Psychology, 57*(1), 136–145. https://doi.org/10.1002/ijop.12805

Vinkers, C. H., van Amelsvoort, T., Bisson, J. I., Branchi, I., Cryan, J. F., Domschke, K., Howes, O. D., Manchia, M., Pinto, L., de Quervain, D., Schmidt, M. V., & van der Wee, N. J. (2020). Stress resilience during the coronavirus pandemic. *European Neuropsychopharmacology, 35*, 12–16. https://doi.org/10.1016/j.euroneuro.2020.05.003

Wade, K. A., Garry, M., Read, J. D., & Lindsay, D. S. (2002). A picture is worth a thousand lies: Using false photographs to create false childhood memories. *Psychonomic Bulletin & Review, 9*(3), 597–603. https://doi.org/10.3758/BF03196318

Wardle, C., & Derakhshan, H. (2017). *Information disorder: Toward an interdisciplinary framework for research and policymaking*. Council of Europe.

Weinstein, N. D. (1980). Unrealistic optimism about future life events. *Journal of Personality and Social Psychology, 39*(5), 806–820. https://doi.org/10.1037/0022-3514.39.5.806

Zarocostas, J. (2020). How to fight an infodemic. *The Lancet, 395*(10225), 676. https://doi.org/10.1016/S0140-6736(20)30461-X

Zepeda, R., Dominguez-Martinez, L., & Silveira, S. (2023). Positive memory recall buffers stress during the COVID-19 pandemic. *Journal of Happiness Studies, 24*(1), 29–48. https://doi.org/10.1007/s10902-022-00604-1

Zhu, B., Chen, C., Loftus, E. F., Lin, C., He, Q., & Chen, C. (2012). Implanted memories of childhood sexual abuse via imagination inflation and confirmatory feedback. *Memory & Cognition, 40*(5), 877–889. https://doi.org/10.3758/s13421-012-0207-7

The Repercussions of the COVID-19 Pandemic on the Contents and Organization of Autobiographical Memory

Öykü Ekinci and Norman R. Brown

Abstract In this chapter, we argue that the COVID-19 pandemic acted as a collective transition and as such affected the contents and organization of autobiographical memory. Although major life transitions are known to affect autobiographical memory organization, the onset of the COVID-19 pandemic presented a unique case that both aligns with and challenges the premises of *Transition Theory*. Despite its global scale and historical significance, the pandemic did not bring about the catastrophic disruptions associated with major collective transitions. Nonetheless, it marked the beginning of a distinct autobiographical period, as evidenced by its frequent use as a temporal landmark and the increased recall of autobiographical memories around its onset. We begin by outlining the concepts within Transition Theory and revisit the theory in light of the recent findings about the pandemic's impact on autobiographical memory. We propose that the pandemic's pervasive presence in media, social environment, and daily life contributed to the changes in how autobiographical memories are organized. We conclude by discussing how this work opens up new questions about the role of collective events in influencing the structure of autobiographical memory.

Keywords Transition Theory · Autobiographical memory · COVID-19 pandemic · COVID bump

Ö. Ekinci (✉) · N. R. Brown
Department of Psychology, University of Alberta, Edmonton, AB, Canada
e-mail: ekinci@ualberta.ca; nrbrown@ualberta.ca

M. Migueles Seco, A. Aizpurua Sanz (eds.), *Autobiographical Memory and the Covid-19 Pandemic*, https://doi.org/10.1007/978-3-032-03895-1_7

1 Introduction

The immediate impacts of the COVID-19 pandemic on daily life, on health, and on society as a whole were undeniable. However, its long-term impact on memory is still coming into focus. In this chapter, we examine the impact of the pandemic on autobiographical memory from a *Transition Theory* perspective (Brown, 2016, 2021). Transition Theory explains how major life transitions affect the organization and content of autobiographical memory by focusing on the changes brought about by such events. In brief, we first introduce the core concepts of Transition Theory and outline its predictions regarding the COVID-19 pandemic. We then present evidence on the pandemic's impact on memory organization and consider the reasons for these effects. We conclude with a discussion concerning the ways the pandemic challenges Transition Theory, highlighting the necessity to consider *thematic pervasiveness* in order to gain a deeper understanding of the relationship between public events and autobiographical memory.

2 Transition Theory and Autobiographical Memory

Transition Theory aims to explain the organization of autobiographical memory by examining how our lives are structured and how life events are experienced, encoded, and interconnected. According to Transition Theory (Brown, 2016), *event components* (ECs) are the building blocks that define and index memorable personal events. On this view, ECs are the identifiable units such as people, places, activities, and objects that are central elements that comprise the event in question. For example, a wedding memory might involve ECs such as the bride and groom, the venue, and specific rituals within the ceremony.

Transition Theory assumes that each experience: (a) forms an *event representation* which consists of a bound set of ECs and (b) acts as a Hebbian learning trial (Hebb, 1949; Lim, 2021; Munakata & Pfaffly, 2004), thereby creating or strengthening connections between co-occurring ECs. These ECs form richly interconnected networks through repeated co-occurrence. For instance, daily routines often involve the same sequence of actions and settings, such as waking up, getting ready in the same environment, and working with the same people. The repeated pairing of these ECs via the Hebbian learning mechanism creates densely interconnected networks that serve as the foundation of lifetime periods.

Major life transitions, such as moving to a new country, play a pivotal role in structuring autobiographical memory. By introducing or terminating large sets of ECs in a coordinated manner, these transitions create distinct boundaries between life periods. For example, after relocating, one's daily life changes dramatically, with old ECs being replaced by new places, people, and activities. This set of synchronized changes creates a clear boundary between life chapters (Thomsen, 2015), and serves to organize memories into distinct periods. In this way, ECs can bear temporal knowledge, helping individuals distinguish between different periods based on their first or last interactions with specific ECs.

In addition to direct experience, post-event processing, such as narrating, ruminating, or discussing events, also forms links between event representations. In addition, as with primary event encodings, these secondary experiences act as learning trials and strengthen connections between co-occurring ECs. Through post-event processing, some event memories are embedded within a broader personal narrative with temporal or causal links to other events, forming *event clusters* (Brown, 2005). Event clusters are groups of memories connected through temporal or causal links (Brown & Schopflocher, 1998a, 1998b; Wright & Nunn, 2000). For instance, the "graduation cluster" might consist of memories of preparation, ceremony, and celebrations that follow. Temporal links are formed from the chronological sequence of events or temporal proximity, such as events that happened within a weekend. Causal links are formed as people (attempt to) understand how and why event sequences unfold in the way that they do (Trabasso & van den Broek, 1985).

Transitions differ widely in their scope, impact, and normativity. Scope refers to the size of the affected group. Some transitions, such as becoming a parent, affect individuals or families, whereas others, such as natural disasters, can drastically change the lives of entire communities. Impact measures the extent of disruption or change caused by the transition. Although some transitions alter nearly all ECs (e.g., immigration), others leave much of daily life intact (e.g., moving houses within the same city). Normativity describes how typical or expected a transition is within a cultural or social framework (Berntsen & Rubin, 2004). For instance, graduating from high school or getting married are considered normative transitions in many societies. In contrast, non-normative transitions, such as accidents or illnesses are typically unexpected (Shi & Brown, 2021).

Transitions affect autobiographical memory organization in several ways. First, major transitions create *bumps* in the temporal distribution of memories, as people tend to report more memories from around the onset of life transitions than from the stable periods that precede and follow them (Schrauf & Rubin, 2001; Shi & Brown, 2016; Shi et al., 2024; Thomsen & Berntsen, 2005). This pattern is driven by the novelty of transitional life events themselves and of the events that are proximate to transitions. Changes in various circumstances lead people to experience many "firsts" and "lasts," and to frequently be involved in unique events. Such events are inherently memorable (Robinson, 1992; Shi et al., 2024) in part because individuals need to spend more cognitive effort to make sense of unfamiliar circumstances (Weiner, 1985; Wong & Weiner, 1981). As a result, this "effort after meaning" increases the salience of these events (Bartlett, 1932; Schrauf & Rubin, 2001; Zaromb & Roediger, 2009). Moreover, the distinctiveness of these transitional and transition-adjacent experiences, combined with the stability of periods that precede and follow major transitions, reduces interference, which also aids in the retention of these memories (Hunt, 2006; Hunt & Rawson, 2011).

Second, important transitions delineate major lifetime periods (Brown et al., 2016; Thomsen & Berntsen, 2008). Looking back on our past, we often define our lives by certain impactful events that encapsulate the circumstances of a particular time. Hearing people refer to their past as "during my college years," "before I met my partner," or "after my child was born" is common. When the transition is public in nature, such as wars, natural disasters, or political upheavals, we refer to these periods as *historically-defined autobiographical periods* (H-DAPs).

The Living-in-History project (Brown et al., 2009) was undertaken to demonstrate that H-DAPs do exist and to identify the conditions under which they emerge. To this end, a standard procedure was developed and used widely. The procedure consists of two main phases. The first involves collecting autobiographical memories through the cue-word method (Crovitz & Shiffman, 1974). During this phase, participants are prompted to recall personal events in response to neutral words (e.g., ball, book, window, box). During the second phase, participants are presented with event memories generated during Phase 1 and are asked to estimate the date of each while thinking out loud.

Because event dating is generally challenging (Brown, 1990; Brown et al., 1986; Friedman, 1993, 2004; Thompson et al., 1993), individuals often use the dates of significant events to draw temporal inferences. Typically, these *temporal landmarks* refer to important lifetime periods (e.g., "while I was in college") or to the transitions that delineate these periods (e.g., "after I moved to New York") (Brown, 2016; Brown et al., 2016; Shum, 1998). On occasion, people use public events or reference periods defined by public events to date personal memories. When this is common, we say that we have observed a *Living-in-History effect* and infer that it reflects the existence of H-DAPs in the sampled population.

Following this line of reasoning, we collected data from over 40 samples across 23 countries (Bohn & Habermas, 2016; Brown et al., 2009, 2012, 2016; Brown & Lee, 2010; Gu et al., 2017; Nourkova & Brown, 2015; Opriş et al., 2023; Zebian & Brown, 2014). In each case, we examined participants' think-aloud protocols for reference to public events and periods. Data obtained from these samples indicates that public events rarely spawn H-DAPs. This is understandable, as it is uncommon to hear someone say, "This happened during the election campaigns" while recalling a mundane memory like bending a spoon while eating ice cream. Our evidence indicates that H-DAPs form only in the wake of cataclysmic public events. For instance, people who experienced the Siege of Sarajevo frequently used this event as a reference point to estimate the dates of other memories (Brown et al., 2009, 2016).

Living-in-History effect appears to be intensity-graded such that it reflects the magnitude of personal change brought about by the event in question. For example, following the Lebanese Civil War, residents of Beirut were twice as likely to use the war as a temporal landmark as were residents of Beqaa Valley—approximately 30 km east of Beirut—where the upheaval was less pronounced (Zebian & Brown, 2014). The same pattern emerged when comparing participants who were rusticated during the Chinese Cultural Revolution with those who were not (Gu et al., 2017), as well as when comparing northern Germans and Berliners for their use of World War II as a temporal landmark (Bohn & Habermas, 2016).

Critically, it appears that it is the change in material circumstances (i.e., ECs) that drives the creation of H-DAPs. We have found that it is neither the psychological impact of the historical event nor its historical or political significance that predicts H-DAP formation. For instance, despite the historical and political importance of the 9/11 attacks, even a sample of New Yorkers did not use this event as a temporal landmark (Brown et al., 2009). Likewise, the fall of Communism in the Soviet Union was rarely mentioned when data were collected from participants in Russia, Azerbaijan, and Uzbekistan (Nourkova & Brown, 2015), Romania (Opriş et al., 2023), or West Germany (Camia et al., 2019). This set of findings underlines the importance of an event's material disruptions on memory organization as opposed to its psychological, symbolic, or historical significance.

3 The COVID-19 Pandemic from the Transition Theory Perspective

The COVID-19 pandemic presents a unique case for Transition Theory. First, the scope of the pandemic was nearly universal, whereas the impact and responses were highly variable between regions. More importantly, unlike typical transitions characterized by the wholesale introduction of a set of new ECs, the pandemic primarily removed familiar ECs without replacing them; Brown (2021) termed this a *transition-by-omission*. Especially during the lockdowns, life was defined by restrictions on activities and social environments. Although certain elements, such as masks and online meetings, were initially novel, they soon became unremarkable (if not annoying) aspects of our daily lives. Although the situation as a whole was unprecedented, our days during the pandemic were often spent with familiar people in familiar places, doing little more than baking sourdough bread and watching Netflix (Heanoy et al., 2022).

For some, the pandemic did cause profound psychological and physiological changes, leaving lasting emotional scars (Li et al., 2023) and shifting perspectives and life strategies (Ausin et al., 2022; Benke et al., 2023; Hannes et al., 2024; Heanoy et al., 2021, 2022). However, for most, except those directly impacted by significant losses or disruptions, the impact of the pandemic was limited. When participants in previous studies were required to rate the degree to which their material circumstances were affected by the pandemic, their responses indicated that they had experienced at most only a modest degree of change (Ekinci & Brown, 2024; Heanoy et al., 2021, 2022). In contrast, period-defining transitions are generally rated high to very high on the Transitional Impact Scale (Enz et al., 2016; Gu et al., 2017; Shi & Brown, 2016; Svob et al., 2014; Thomsen et al., 2021). Furthermore, because life has largely returned to normal (Gallup, 2023, 2024; Statista, 2023), the pandemic may not have resulted in enduring changes for many people.

The unusual nature of the pandemic, which encompassed both stability and change, and both novelty and routine, raises the question of whether the pandemic gave rise to a *COVID-19 H-DAP*, or whether it is represented for most merely as an *extended interlude.* In Transition Theory, an interlude is defined as a period of time during which people experience a radical, but temporary, disruption in the fabric of daily life (Brown, 2021). Note, that people experience both positive interludes (e.g., a 2-week vacation in Tulum) and negative interludes (e.g., a 2-week evacuation following a flood). Regardless, the important point is that after the interlude, life is more or less the same as life before the interlude.

From the Transition Theory perspective, the start of the lockdowns profoundly disrupted the usual pattern of daily life; it did so by introducing a small number of new ECs and limiting access to a large number of familiar ones. Unlike other major transitions, which are generally followed by an eventful adjustment phase, the period after the pandemic onset was repetitive and quickly became mundane. Nonetheless, in the "background," COVID-19 was omnipresent as it dominated mass media, public discourse, and interpersonal communication for almost 2 years (Okada et al., 2024; Pew Research Center, 2022; Wirz et al., 2022). This suggests that secondary processing of experiences, such as thinking and talking about it, reflecting and ruminating about the events, was heightened during this period and may have amplified the mnemonic salience of the pandemic.

4 COVID-19's Mnemonic Impact

In a theoretical article published in 2021, Brown considered the possibility that the pandemic had served as a major collective transition and that as such it fostered the formation of a new widespread H-DAP—the COVID period. As noted above, major transitions produce bumps—an increase in memorable events around the time of transition. Thus, if the pandemic were experienced as a major transition, there should be evidence for a *COVID-bump,* an increase in the memories from around the onset of the pandemic. Given the monotonous nature of the pandemic, it also seemed reasonable to predict a normalization effect. The idea here was that people would soon schematize the novel aspects of the pandemic and that COVID restrictions would limit people's abilities to have novel, potentially interesting experiences. Taken together, these facts imply that people should have access to fewer autobiographical memories from 2021 than from 2020.

Evidence supporting these predictions comes, in part, from a large sample assessed multiple times in 2020 and the following years. Rouhani et al. (2023) asked participants to describe 2020 in terms of personally important events, experiences, and feelings and allowed them to provide as many entries as they could. The temporal distribution of these entries created a distinct bump when participants were asked to recall 2020 at the end of 2020, during 2021, and during 2022. Around 25% of the entries were from March 2020, while the remaining entries were more or less evenly distributed across the remaining months. Notably, when recalling 2021, there was no bump for any of the months, highlighting the fact the March 2020 bump was due to the onset of the pandemic. Although methodological considerations warrant caution (e.g., overly broad definition of "memory"), the results align with the Transition

Theory predictions, showing both a COVID-bump and subsequent normalization of pandemic-related circumstances by the following year.

This normalization effect was further supported by Topçu and Hirst (2024), who compared memories and future projections for personal and public events. Importantly, these researchers asked participants to indicate whether the events that were recalled/projected were directly related to the pandemic, somewhat related, or unrelated to it. In the first data collection conducted in 2020, 64% of the responses were COVID-related, but this dropped to 33% in subsequent data collection conducted in 2021. These results provide additional evidence for the waning salience of the pandemic over time.

Our lab has focused on the pandemic's position as an H-DAP by investigating the use of the COVID-19 pandemic as a temporal landmark in dating personal events. We collected data from first-year undergraduate students, a group that had recently experienced major transitional events, such as graduating high school and starting university, around the same time. The participants were asked to recall 12 memories in response to neutral cue words, and then to date the events thinking aloud (see Ekinci & Brown, 2024 for more details). They then indicated whether each event was related, indirectly related, or unrelated to the pandemic and rated memory and event characteristics such as importance, vividness, and emotional valence. Finally, they answered questions on the magnitude of the pandemic's impact on their material and psychological circumstances using the Transitional Impact Scale (Svob et al., 2014).

The time frame to recall memories was limited to 16 months, so that we could capture the month-to-month variations in event references. In this way, it differed from previous Living-in-History studies which used longer timeframes. The study involved two samples tested at different times. The first group was tested in the spring of 2021 and recalled events from September 2019 to December 2020 range. The second group was tested in fall 2021 and recalled events from the beginning of 2020 to April 2021. Notably, the first group's memories could include events from up to 6 months prior to the onset of the pandemic, whereas the second group's timeframe only allowed for a 2-month pre-pandemic window. We aimed to investigate whether references to the pandemic would persist even when the pre-pandemic period was largely excluded.

Our results underlined the ubiquity of the COVID-19 pandemic in terms of the conditions it created and its implications for autobiographical memory organization. Consistent with our argument that the pandemic did not produce calamitous and enduring material changes, the participants rated the change in material circumstances as moderate; they also indicated that they experienced more psychological change than material change. In both samples, around one-third of memories were considered COVID-related by the participants, with a drop in the second group (from 39.1% to 31.3%) when data were collected further into the pandemic. The recall period for the first group began in September 2019, while the second group reported memories starting in January 2020. As a result, the first group had 6 months without the pandemic, while the second group had only 2 months. Despite this

Table 1 A summary of mean ratings and standard deviations for the memory characteristics

COVID-relation	Importance *M (SD)*	Vividness *M (SD)*	Self-relevance *M (SD)*	Emotional valence *M (SD)*
Unrelated	2.26 (1.27)	3.73 (1.14)	1.99 (1.15)	0.27 (1.08)
Indirectly related	2.52 (1.36)	4.11 (0.86)	2.15 (1.22)	0.53 (1.10)
Directly related	2.87 (1.47)	4.37 (0.83)	2.65 (1.40)	0.48 (1.17)

Note: Means and standard deviations are presented for each level of relatedness, averaged across both waves of data collection. Ratings for importance, vividness, and self-relevance range from 1 to 5, with lower scores indicating a lower value for the characteristic. Emotional valence ratings range from −2 (extremely negative) to +2 (extremely positive), with scores around 0 indicating neutral memories

difference, the percentage of COVID-related memories still decreased, further supporting the idea of a normalization effect for the COVID-19 pandemic.

Across both waves of data collection, memories directly related to the pandemic were rated higher in importance, vividness, and self-relevance compared to unrelated memories (see Table 1). Indirectly related memories usually fall in between. However, even for COVID-related memories, the mean ratings were often significantly below the midpoint of the five-point scale for importance and self-relevance. In contrast, memories were evaluated to be highly vivid in general. Consistent with prior research (e.g., Shi et al., 2024), these data indicate that event memories recalled in this study were vivid and varied in their relative importance and relation to self, but were not seen as highly significant overall. Similarly, emotional valence ratings, rated on a scale from −2 to 2 ranging from very negative to very positive, did not significantly differ between the three groups of memories in either data collection wave. The mean ratings fell between 0 and 1, indicating these memories were overall slightly positive. This suggests that the memories collected were not necessarily central to individuals' lives or emotionally intense but rather more typical autobiographical memories from everyday life, regardless of their connection to the pandemic.

The analysis of event dating protocols revealed that the COVID-19 pandemic served as an important temporal landmark (in 16.8% of memories in Wave 1 and 16.1% in Wave 2). This frequency is comparable to other recent major life periods, such as high school and university. Although memories that had direct or indirect relevance to the pandemic were more likely to be dated with reference to COVID-19, there was also a notable portion of unrelated memories that were dated with a COVID-reference (13.7% in Wave 1 and 9.2% in Wave 2).

Event dating references to the pandemic were most frequently used for events around its onset in March 2020. Similarly, references to high school peaked around the start of the senior year and graduation, while university references were most commonly used for memories around the beginning of the academic year. Other types of references, such as personal events or general weather conditions, were more evenly distributed throughout the year. The graded nature of these references reflects the fact that people are more likely to use these life periods as reference points when dating events close to the transition that initiated the period. This supports the prominence of transitions in organizing autobiographical memory.

The results from Ekinci & Brown (2024) indicate that the pandemic can be characterized as an H-DAP, suggesting that it was a major transition that affected how autobiographical memories are organized. This finding contrasts with the evidence from past studies that demonstrated that H-DAPs arise only when an event has had catastrophic effects on individuals' lives. Participant assessments of the material impact of the pandemic, and the decline in mean ratings as we move further from the onset of the pandemic (Ekinci & Brown, 2024; Heanoy et al., 2021, 2022) are a reflection of the pandemic's moderate and impermanent effects. Typically, such events, regardless of their historical significance, do not lead to a Living-in-History effect (Brown et al., 2009; Brown & Lee, 2010; Nourkova & Brown, 2015; Opriş et al., 2023).

5 Rethinking Transition Theory

Unlike other public events involving abrupt, tangible, and enduring changes in daily life, the COVID-19 pandemic's effects were primarily psychological and thematic and pervade mental, social, and media landscapes rather than directly altering material conditions for most people. Brown (2021) proposed that the pervasive nature of the pandemic may have disrupted typical memory processes involved in H-DAP formation. The pandemic's widespread presence in every context may have made "life during the pandemic" a central reference point in memory, even for individuals who experienced minimal direct disruption.

A defining characteristic of the pandemic was the intensification of information flow, which potentially influenced how people remembered this period. Research shows a rapid increase in online discussions about COVID-19 during its early stages (Ashokkumar & Pennebaker, 2021; Britt et al., 2021), alongside substantial news media coverage (Krawczyk et al., 2021; Pew Research Center, 2020a, 2022; Rauchfleisch et al., 2023; Wirz et al., 2022), and extensive interpersonal conversations (Okada et al., 2024; Wagner & Reifegerste, 2023). National news sources became the primary source of information, and COVID-19 news attracted high levels of public attention (Pew Research Center, 2020b). Pandemic-related discussions extended beyond media consumption. Individuals frequently discussed pandemic news and experiences with their family, friends, partners, and colleagues (Link et al., 2021). The saturation of information and conversations around it likely amplified the pandemic's pervasiveness, created a collective narrative, and provided a collective temporal framework; even for those less directly affected by the pandemic.

This intense focus on the pandemic aligns with linguistic social media analyses that demonstrated a surge in cognitive-processing words, terms indicating self-reflection, causation, and uncertainty, at the outset of the pandemic (Ashokkumar & Pennebaker, 2021). Notably, although similar patterns arise following other public crises, such as 9/11 (Cohn et al., 2004) and natural disasters (Correa et al., 2016; Pennebaker & Harber, 1993), the persistence of this trend during the pandemic was unprecedented. Online conversations that reflect high levels of engagement with the pandemic continued for months after the initial outbreak. This further underlines the pandemic's sustained influence on the collective discourse. The COVID-19 pandemic thus represents a unique form of collective transition that affected the organization of autobiographical memory through its pervasive repercussions rather than material changes.

6 Conclusion

The COVID-19 pandemic stands out as a unique case of global disruption. Although the pandemic lacked the destructive impact often associated with historically defined autobiographical periods, it nevertheless affected the organization of autobiographical memory. This conclusion motivated us to revise Transition Theory, as one of its main assumptions was challenged by the COVID-19 pandemic. The pandemic displayed H-DAP characteristics, not just for those undergoing extreme disruptions but for a broad range of individuals whose daily lives were altered in smaller ways for a limited period of time. The presence of COVID-19 as a dominant theme in social environments and media spaces created a shared, albeit different, experience and unusual conditions that were reflected in the contents and organization of autobiographical memory. This raises the question of whether there are other local collective events on a smaller scale that infiltrate into people's lives without overturning those lives yet still give rise to identifiable autobiographical periods.

For some individuals, the COVID-19 pandemic significantly disrupted their lives; this is particularly true for frontline workers and those who experienced irrevocable losses (Ashcroft et al., 2022; Shaukat et al., 2020). In these cases, the Living-in-History effect would likely be more pronounced because the pandemic brought on marked changes in life circumstances. Similar patterns have been observed in other contexts, such as the Balkans (Brown et al., 2009), Lebanon (Zebian & Brown, 2014), Germany (Bohn & Habermas, 2016), China (Gu et al., 2017), and Bangladesh (Islam & Haque, 2022), where the Living-in-History effect was stronger for those who underwent greater material changes compared to those who did not. Although factors like age and gender do not appear to influence this effect, future research could examine whether certain populations are more likely to organize autobiographical memory around collective transitions based on the severity of their material or psychological impact.

For students, the strangeness of online schooling during their formative years, along with disruptions to social and extracurricular activities, might have created vivid memories that could contribute to the Living-in-History effect in the long term. These changes, although not as severe as those experienced by individuals facing direct and irrevocable losses, still marked a major shift in how they lived and how they interacted for an extended period of time leading in some cases to last effects on their mental health, their learning experiences and the content of the events they will remember (Felfe et al., 2023; Hjuler et al., 2025; Mazrekaj & De Witte, 2024).

Looking ahead, it is worth considering whether the prominence of pandemic-related memories and references will reduce over time or whether the COVID-19 pandemic will remain a major life period. The current salience of the pandemic as a temporal landmark is clear in our findings. However, it is possible that its influence may wane as individuals go through new life transitions or as collective attention shifts. Alternatively, it is also possible that the pandemic will endure as an H-DAP due to its unprecedented global impact and the way it affected our lives. Future studies revisiting these questions over the next decade or more will be important in understanding whether it becomes a lasting point of reference or gradually fades into the background of autobiographical memories.

References

Ashcroft, R., Sur, D., Greenblatt, A., & Donahue, P. (2022). The impact of the COVID-19 pandemic on social workers at the frontline: A survey of Canadian social workers. *The British Journal of Social Work, 52*(3), 1724–1746. https://doi.org/10.1093/bjsw/bcab158

Ashokkumar, A., & Pennebaker, J. W. (2021). Social media conversations reveal large psychological shifts caused by COVID-19's onset across US cities. *Science Advances, 7*(39), Article eabg7843. https://doi.org/10.1126/sciadv.abg7843. https://doi.org/10.1126/sciadv.abg7843

Ausín, B., González-Sanguino, C., Castellanos, M. Á., Sáiz, J., Zamorano, S., Vaquero, C., & Muñoz, M. (2022). The psychological impact of the COVID-19 pandemic in Spain: A longitudinal study. *Psicothema, 34*(1), 66–73. https://doi.org/10.7334/psicothema2021.290

Bartlett, F. C. (1932). *Remembering: A study in experimental and social psychology*. Cambridge University Press.

Benke, C., Autenrieth, L. K., Asselmann, E., & Pané-Farré, C. A. (2023). One year after the COVID-19 outbreak in Germany: Long-term changes in depression, anxiety, loneliness, distress and life satisfaction. *European Archives of Psychiatry and Clinical Neuroscience, 273*(2), 289–299. https://doi.org/10.1007/s00406-022-01400-0

Berntsen, D., & Rubin, D. C. (2004). Cultural life scripts structure recall from autobiographical memory. *Memory & Cognition, 32*(3), 427–442. https://doi.org/10.3758/BF03195836

Bohn, A., & Habermas, T. (2016). Living in history and living by the cultural life script: How older Germans date their autobiographical memories. *Memory, 24*(4), 482–495. https://doi.org/10.1080/09658211.2015.1019890

Britt, R. K., Britt, B. C., Panek, E., & Lee, J. (2021). Communication expressed on the COVID-19 subreddit in the midst of a global pandemic. *Health Communication, 38*(6), 1157–1167. https://doi.org/10.1080/10410236.2021.1994190

Brown, N. R. (1990). Organization of public events in long-term memory. *Journal of Experimental Psychology: General, 119*(3), 297–314. https://doi.org/10.1037/0096-3445.119.3.297

Brown, N. R. (2005). On the prevalence of event clusters in autobiographical memory. *Social Cognition, 23*(1), 35–69. https://doi.org/10.1521/soco.23.1.35.59194

Brown, N. R. (2016). Transition theory: A minimalist perspective on the organization of autobiographical memory. *Journal of Applied Research in Memory and Cognition, 5*(2), 128–134. https://doi.org/10.1016/j.jarmac.2016.03.005

Brown, N. R. (2021). The possible effects of the COVID-19 pandemic on the contents and organization of autobiographical memory: A Transition-Theory perspective. *Cognition, 212*, Article 104694. https://doi.org/10.1016/j.cognition.2021.104694

Brown, N. R., Hansen, T. G. B., Lee, P. J., Vanderveen, S. A., & Conrad, F. G. (2012). Historically defined autobiographical periods: Their origins and implications. In D. Berntsen & D. C. Rubin (Eds.), *Understanding autobiographical memory: Theories and approaches* (pp. 160–180). Cambridge University Press. https://doi.org/10.1017/CBO9781139021937.013

Brown, N. R., & Lee, P. J. (2010). Public events and the organization of autobiographical memory: An overview of the Living-in-History Project. *Behavioral Sciences of Terrorism and Political Aggression, 2*(2), 133–149. https://doi.org/10.1080/19434471003597431

Brown, N. R., Lee, P. J., Krslak, M., Conrad, F. G., Hansen, T. G. B., Havelka, J., & Reddon, J. R. (2009). Living in history: How war, terrorism, and natural disaster affect the organization of autobiographical memory. *Psychological Science, 20*(4), 399–405. https://doi.org/10.1111/j.1467-9280.2009.02307.x

Brown, N. R., & Schopflocher, D. (1998a). Event clusters: An organization of personal events in long-term memory. *Psychological Science, 9*(6), 470–475. https://doi.org/10.1111/1467-9280.00087

Brown, N. R., & Schopflocher, D. (1998b). Event cueing, event clusters, and the temporal distribution of autobiographical memories. *Applied Cognitive Psychology, 12*(4), 305–319. https://doi.org/10.1002/(SICI)1099-0720(199808)12:4%3C305::AID-ACP569%3E3.0.CO;2-5

Brown, N. R., Schweickart, O., & Svob, C. (2016). The effect of collective transitions on the organization and contents of autobiographical memory: A transition theory perspective. *The American Journal of Psychology, 129*(3), 259–282. https://doi.org/10.5406/amerjpsyc.129.3.0259

Brown, N. R., Shevell, S. K., & Rips, L. J. (1986). Public events and their personal context. In D. C. Rubin (Ed.), *Autobiographical memory* (pp. 137–158). Cambridge University Press. https://doi.org/10.1017/CBO9780511558313.014

Camia, C., Menzel, C., & Bohn, A. (2019). A positive Living-in-History effect: The case of the fall of the Berlin Wall. *Memory, 27*(10), 1381–1389. https://doi.org/10.1080/09658211.2019.1661494

Cohn, M. A., Mehl, M. R., & Pennebaker, J. W. (2004). Linguistic markers of psychological change surrounding September 11, 2001. *Psychological Science, 15*(10), 687–693. https://doi.org/10.1111/j.0956-7976.2004.00741.x

Correa, T., Scherman, A., & Arriagada, A. (2016). Audiences and disasters: Analyses of media diaries before and after an earthquake and a massive fire. *Journal of Communication, 66*(4), 519–541. https://doi.org/10.1111/jcom.12245

Crovitz, H. F., & Schiffman, H. (1974). Frequency of episodic memories as a function of their age. *Bulletin of the Psychonomic Society, 4*(5-B), 517–518. https://doi.org/10.3758/BF03334277

Ekinci, Ö., & Brown, N. R. (2024). The COVID-19 pandemic as autobiographical period: Evidence from an event dating study. *Memory, 32*(2), 283–291. https://doi.org/10.1080/09658211.2024.2310562

Enz, K. F., Pillemer, D. B., & Johnson, K. M. (2016). The relocation bump: Memories of middle adulthood are organized around residential moves. *Journal of Experimental Psychology: General, 145*(8), 935–940. https://doi.org/10.1037/xge0000188

Felfe, C., Saurer, J., Schneider, P., Vornberger, J., Erhart, M., Kaman, A., & Ravens-Sieberer, U. (2023). The youth mental health crisis: Quasi-experimental evidence on the role of school closures. *Science Advances, 9*(33), eadh4030. https://doi.org/10.1126/sciadv.adh4030

Friedman, W. J. (1993). Memory for the time of past events. *Psychological Bulletin, 113*(1), 44–66. https://doi.org/10.1037/0033-2909.113.1.44

Friedman, W. J. (2004). Time in autobiographical memory. *Social Cognition, 22*(5), 591–605. https://doi.org/10.1521/soco.22.5.591.50766

Gallup. (2023, October 16). *Roundup of Gallup's COVID-19 coverage.* https://news.gallup.com/opinion/gallup/308126/roundup-gallup-covid-coverage.aspx

Gallup. (2024, March 18). *Four years later, 59% say COVID-19 pandemic is over.* https://news.gallup.com/poll/612230/four-years-say-covid-pandemic.aspx

Gu, X., Tse, C. S., & Brown, N. R. (2017). The effects of collective and personal transitions on the organization and contents of autobiographical memory in older Chinese adults. *Memory & Cognition, 45*(8), 1335–1349. https://doi.org/10.3758/s13421-017-0733-0

Hannes, C., Schiffer, S., & von Nitzsch, R. (2024). Changes in value priorities due to the COVID-19 pandemic—A 4-year cross-sectional study with German students. *PLoS One, 19*(1), Article e0297236. https://doi.org/10.1371/journal.pone.0297236

Heanoy, E. Z., Shi, L., & Brown, N. R. (2021). Assessing the transitional impact and mental health consequences of the Covid-19 pandemic onset. *Frontiers in Psychology, 11*, Article 607976. https://doi.org/10.3389/fpsyg.2020.607976

Heanoy, E. Z., Uzer, T., & Brown, N. R. (2022). COVID-19 pandemic as a transitional event: From the perspective of the transition theory. *Encyclopedia, 2*(3), 1602–1610. https://doi.org/10.3390/encyclopedia2030109

Hebb, D. O. (1949). *The organization of behavior: A neuropsychological theory*. Wiley.

Hjuler, T. F., Lee, D., & Ghetti, S. (2025). Remembering history: Autobiographical memory for the COVID-19 pandemic lockdowns, psychological adjustment, and their relation over time. *Child Development, 96*(1), 55–70. https://doi.org/10.1111/cdev.14131

Hunt, R. R. (2006). The concept of distinctiveness in memory research. In R. R. Hunt & J. B. Worthen (Eds.), *Distinctiveness and memory* (pp. 3–26). https://doi.org/10.1093/acprof:oso/9780195169669.003.0001

Hunt, R. R., & Rawson, K. A. (2011). Knowledge affords distinctive processing in memory. *Journal of Memory and Language, 65*(4), 390–405. https://doi.org/10.1016/j.jml.2011.07.001

Islam, A., & Haque, S. (2022). Living-in-history effect in the dating of important autobiographical memories. *Memory & Cognition, 50*(5), 1078–1089. https://doi.org/10.3758/s13421-021-01250-6

Krawczyk, K., Chelkowski, T., Laydon, D. J., Mishra, S., Xifara, D., Gibert, B., Flaxman, S., Mellan, T., Schwämmle, V., Röttger, R., Hadsund, J. T., & Bhatt, S. (2021). Quantifying online news media coverage of the COVID-19 pandemic: Text mining study and resource. *Journal of Medical Internet Research, 23*(6), Article e28253. https://doi.org/10.2196/28253

Li, Z., He, J., Wang, Y., Bai, M., Zhang, Y., Chen, H., Li, W., Cai, Y., Chen, S., Qu, M., & Wang, J. (2023). A cross-sectional study on the mental health of patients with COVID-19 1 year after discharge in Huanggang, China. *European Archives of Psychiatry and Clinical Neuroscience, 273*(2), 301–310. https://doi.org/10.1007/s00406-022-01484-8

Lim, S. (2021). Hebbian learning revisited and its inference underlying cognitive function. *Current Opinion in Behavioral Sciences, 38*, 96–102. https://doi.org/10.1016/j.cobeha.2021.02.006

Link, E., Rosset, M., & Freytag, A. (2021). Typen des interpersonalen und medialen Kommunikations-und Informationshandelns während der Corona-Pandemie. In F. Sukalla & C. Voigt (Eds.), *Risiken und Potenziale in der Gesundheitskommunikation* (pp. 21–30). Deutsche Gesellschaft für Publizistik- und Kommunikationswissenschaft. https://doi.org/10.21241/ssoar.74286

Mazrekaj, D., & De Witte, K. (2024). The impact of school closures on learning and mental health of children: Lessons from the COVID-19 pandemic. Perspectives on Psychological Science, 19(4), 686-693. https://doi.org/10.1177/17456916231181108

Munakata, Y., & Pfaffly, J. (2004). Hebbian learning and development. *Developmental Science, 7*(2), 141–148. https://doi.org/10.1111/j.1467-7687.2004.00331.x

Nourkova, V. V., & Brown, N. R. (2015). Assessing the impact of "the collapse" on the organization and content of autobiographical memory in the former Soviet Union. *Journal of Social Issues, 71*(2), 324–337. https://doi.org/10.1111/josi.12113

Okada, H., Okuhara, T., & Kiuchi, T. (2024). The direct effects of media exposure on behaviors aimed at preventing COVID-19 and its indirect effects as mediated by interpersonal communication: A longitudinal study in Japan. *Frontiers in Public Health, 12*, Article 1454978. https://doi.org/10.3389/fpubh.2024.1454978

Opriş, A. M., Visu-Petra, L., & Brown, N. R. (2023). Living in history and by the cultural life script: What events modulate autobiographical memory organization in a sample of older adults from Romania? *Memory Studies, 16*(4), 912–927. https://doi.org/10.1177/17506980221108476

Pennebaker, J. W., & Harber, K. (1993). A social stage model of collective coping: The Persian Gulf and other natural disasters. *Journal of Social Issues, 49*(4), 125–145. https://doi.org/10.1111/j.1540-4560.1993.tb01184.x

Pew Research Center. (2020a, May 1). *Most Americans say COVID-19 has changed news reporting; Many unsure how it's affected the industry.* https://www.pewresearch.org/short-reads/2020/05/01/most-americans-say-covid-19-has-changed-news-reporting-but-many-are-unsure-how-its-affected-the-industry/

Pew Research Center. (2020b, May 20). *Americans who rely most on White House for COVID-19 news more likely to downplay the pandemic.* https://www.pewresearch.org/journalism/2020/05/20/americans-who-rely-most-on-white-house-for-covid-19-news-more-likely-to-downplay-the-pandemic/

Pew Research Center. (2022, January 27). *Attention to COVID-19 news increased slightly amid omicron surge; partisans differ in views about the outbreak.* https://www.pewresearch.org/short-reads/2022/01/27/attention-to-covid-19-news-increased-slightly-amid-omicron-surge-partisans-differ-in-views-about-the-outbreak/

Rauchfleisch, A., Siegen, D., & Vogler, D. (2023). How COVID-19 displaced climate change: Mediated climate change activism and issue attention in the Swiss media and online sphere. *Environmental Communication, 17*(3), 313–321. https://doi.org/10.1080/17524032.2021.1990978

Robinson, J. A. (1992). First experience memories: Contexts and functions in personal histories. In M. A. Conway, D. C. Rubin, H. Spinnler, & W. A. Wagenaar (Eds.), *Theoretical perspectives on autobiographical memory* (pp. 233–239). Springer. https://doi.org/10.1007/978-94-015-7967-4_13

Rouhani, N., Stanley, D., COVID-Dynamic Team, & Adolphs, R. (2023). Collective events and individual affect shape autobiographical memory. *Proceedings of the National Academy of Sciences, 120*(29), Article e2221919120. https://doi.org/10.1073/pnas.2221919120

Schrauf, R. W., & Rubin, D. C. (2001). Effects of voluntary immigration on the distribution of autobiographical memory over the lifespan. *Applied Cognitive Psychology, 15*(7), S75–S88. https://doi.org/10.1002/acp.835

Shaukat, N., Ali, D. M., & Razzak, J. (2020). Physical and mental health impacts of COVID-19 on healthcare workers: A scoping review. *International Journal of Emergency Medicine, 13*, Article 40. https://doi.org/10.1186/s12245-020-00299-5

Shi, L., & Brown, N. R. (2016). The effect of immigration on the contents and organization of autobiographical memory: A transition-theory perspective. *Journal of Applied Research in Memory and Cognition, 5*(2), 135–142. https://doi.org/10.1016/j.jarmac.2016.03.002

Shi, L., & Brown, N. R. (2021). Beliefs about transitional events: The effect of experience and life-script consistency. *Frontiers in Psychology, 12*, Article 727524. https://doi.org/10.3389/fpsyg.2021.727524

Shi, L., Brown, N. R., & Reimer, P. C. (2024). Exploring the metamnemonic and phenomenal differences between transitional and mundane events. *Memory & Cognition, 52*(5), 1164–1181. https://doi.org/10.3758/s13421-024-01534-7

Shum, M. S. (1998). The role of temporal landmarks in autobiographical memory processes. *Psychological Bulletin, 124*(3), 423–442. https://doi.org/10.1037/0033-2909.124.3.423

Statista. (2023, November 7). *The Covid jobs recovery is complete (And then some).* https://www.statista.com/chart/24363/number-of-jobs-lost-since-february-2020/

Svob, C., Brown, N. R., Reddon, J. R., Uzer, T., & Lee, P. J. (2014). The transitional impact scale: Assessing the material and psychological impact of life transitions. *Behavior Research Methods, 46*(2), 448–455. https://doi.org/10.3758/s13428-013-0378-2

Thompson, C. P., Skowronski, J. J., & Betz, A. L. (1993). The use of partial temporal information in dating personal events. *Memory & Cognition, 21*(3), 352–360. https://doi.org/10.3758/bf03208267

Thomsen, D. K. (2015). Autobiographical periods: A review and central components of a theory. *Review of General Psychology, 19*(3), 294–310. https://doi.org/10.1037/gpr0000043

Thomsen, D. K., & Berntsen, D. (2005). The end point effect in autobiographical memory: More than a calendar is needed. *Memory, 13*(8), 846–861. https://doi.org/10.1080/09658210444000449

Thomsen, D. K., & Berntsen, D. (2008). The cultural life script and life story chapters contribute to the reminiscence bump. *Memory, 16*(4), 420–435. https://doi.org/10.1080/09658210802010497

Thomsen, D. K., Talarico, J. M., & Steiner, K. L. (2021). When does a wedding mark the beginning of a new chapter in one's life? *Scandinavian Journal of Psychology, 62*(5), 675–682. https://doi.org/10.1111/sjop.12754

Topçu, M. N., & Hirst, W. (2024). When the personal and the collective intersects: Memory, future thinking, and perceived agency during the COVID-19 pandemic. *Journal of Experimental Psychology: General, 153*(9), 2258–2278. https://doi.org/10.1037/xge0001624

Trabasso, T., & van den Broek, P. (1985). Causal thinking and the representation of narrative events. *Journal of Memory and Language, 24*(5), 612–630. https://doi.org/10.1016/0749-596X(85)90049-X

Wagner, A., & Reifegerste, D. (2023). "The part played by people" in times of COVID-19: Interpersonal communication about media coverage in a pandemic crisis. *Health Communication, 38*(5), 1014–1021. https://doi.org/10.1080/10410236.2021.1989786

Weiner, B. (1985). "Spontaneous" causal thinking. *Psychological Bulletin, 97*(1), 74–84. https://doi.org/10.1037/0033-2909.97.1.74

Wirz, C. D., Shao, A., Bao, L., Howell, E. L., Monroe, H., & Chen, K. (2022). Media systems and attention cycles: Volume and topics of news coverage on COVID-19 in the United States and China. *Journalism & Mass Communication Quarterly, 99*(4), 1048–1071. https://doi.org/10.1177/10776990211049455

Wong, P. T., & Weiner, B. (1981). When people ask "why" questions, and the heuristics of attributional search. *Journal of Personality and Social Psychology, 40*(4), 650–663. https://doi.org/10.1037/0022-3514.40.4.650

Wright, D. B., & Nunn, J. A. (2000). Similarities within event clusters in autobiographical memory. *Applied Cognitive Psychology, 14*(5), 479–489. https://doi.org/10.1002/1099-0720(200009)14:5<479::AID-ACP688>3.0.CO;2-C

Zaromb, F. M., & Roediger, H. L. (2009). The effects of "effort after meaning" on recall: Differences in within- and between-subjects designs. *Memory & Cognition, 37*(4), 447–463. https://doi.org/10.3758/MC.37.4.447

Zebian, S., & Brown, N. R. (2014). Living in history in Lebanon: The influence of chronic social upheaval on the organisation of autobiographical memories. *Memory, 22*(3), 194–211. https://doi.org/10.1080/09658211.2013.775310

Index

M. Migueles Seco, A. Aizpurua Sanz (eds.), *Autobiographical Memory and the Covid-19 Pandemic*, https://doi.org/10.1007/978-3-032-03895-1

GPSR Compliance

The European Union's (EU) General Product Safety Regulation (GPSR) is a set of rules that requires consumer products to be safe and our obligations to ensure this.

If you have any concerns about our products, you can contact us on ProductSafety@springernature.com

In case Publisher is established outside the EU, the EU authorized representative is:

Springer Nature Customer Service Center GmbH
Europaplatz 3
69115 Heidelberg, Germany

Batch number: 09457522

Printed by Printforce, the Netherlands